RAND ARROYO CENTER

Are U.S. Military Interventions Contagious over Time?

Intervention Timing and Its Implications for Force Planning

Jennifer Kavanagh

Prepared for the United States Army

The research described in this report was sponsored by the United States Army under Contract No. W74V8H-06-C-0001.

Library of Congress Cataloging-in-Publication Data

Kavanagh, Jennifer, 1981-
  Are U.S. military interventions contagious over time? : intervention timing and its implication for force planning / Jennifer Kavanagh.
    p cm
  Includes bibliographical references.
  ISBN 978-0-8330-7901-5 (pbk. : alk. paper)
  1. Military planning—United States. 2. Intervention (International law) 3. United States—Military policy—Case studies. 4. Intervention (International law)—Case studies. I. Title.

U153.K38 2013
355'.033573—dc23   2013016406

The RAND Corporation is a nonprofit institution that helps improve policy and decisionmaking through research and analysis. RAND's publications do not necessarily reflect the opinions of its research clients and sponsors.

RAND® is a registered trademark.

Published 2013 by the RAND Corporation
1776 Main Street, P.O. Box 2138, Santa Monica, CA 90407-2138
1200 South Hayes Street, Arlington, VA 22202-5050
4570 Fifth Avenue, Suite 600, Pittsburgh, PA 15213-2665
RAND URL: http://www.rand.org/
To order RAND documents or to obtain additional information, contact
Distribution Services: Telephone: (310) 451-7002;
Fax: (310) 451-6915; Email: order@rand.org

Preface

This report documents the results of one task supporting the project "Improving Data Collection to Inform Future Planning." The task considered how data could be used more effectively to inform long-range Army force planning. The report should be of interest to force planners across the defense analytical community, particularly those involved in applying quantitative techniques to long-range planning challenges. The information cutoff date for this document is September 2011.

This research was sponsored by the Army Quadrennial Defense Review Office in the Office of the Deputy Chief of Staff, G-8, Headquarters, Department of the Army, and was conducted within RAND Arroyo Center's Strategy, Doctrine, and Resources Program. RAND Arroyo Center, part of the RAND Corporation, is a federally funded research and development center sponsored by the United States Army.

The Project Unique Identification Code (PUIC) for the project that produced this document is HQD115824.

For more information on RAND Arroyo Center, contact the Director of Operations (telephone 310-393-0411, extension 6419; fax 310-451-6952; email Marcy_Agmon@rand.org) or visit Arroyo's website at http://www.rand.org/ard/.

Contents

Preface iii
Figures vii
Tables ix
Summary xi
Acknowledgements xv
Abbreviations xvii

CHAPTER ONE
Introduction 1

CHAPTER TWO
Defining Temporal Dependence: A Review of Existing Evidence 5
What Is Temporal Dependence? 5
What Does the Literature Say About Intervention Timing and Temporal Dependence? 6
Interventions and Timing 6
Predictors of Armed Conflict and Political Instability 8
Temporal Dependence in Financial Markets 10
Summary 11

CHAPTER THREE
Testing for Temporal Dependence 13
Methodology 13
Data and Operationalization 15
   Interventions 15
   Armed Conflict 19
Results 22
What Drives Armed Conflict? 25
   Testing for Robustness: Linear and ARIMA Specifications 28
   Summary 32
Is There Temporal Dependence Between Military Deployments? 33
   Testing for Robustness: Linear and ARIMA Specifications 38
   Summary 42

CHAPTER FOUR
Implications for Force Planning 45
Will Temporal Dependence Affect Force Requirements? 45
Mechanisms of Temporal Dependence 47
How Can Temporal Dependence Be Integrated into the Planning Process? 48
    Assessing the Relevance of Temporal Clustering 49
    Building Temporal Dependence into Force Planning 49
    Avoiding Clustered Interventions 51

CHAPTER FIVE
Conclusion and Next Steps 53

Bibliography 55

Figures

3.1. Interventions, by Year of Onset 16
3.2. Armed Conflict, Onset 19
3.3. Armed Conflict, Cumulative 20

Tables

3.1. Poisson Models, Incidence Ratios, Temporal Clustering of Conflicts, Models 1–5.1 23

3.2. Poisson Models, Incidence Ratios, Temporal Clustering of Conflicts, Models 6–16 24

3.3. Linear Models, Temporal Clustering of Conflicts, Models 1a–5.1a 28

3.4. Linear Models, Temporal Clustering of Conflicts, Models 6a–14a 29

3.5. ARIMA Models, Temporal Clustering of Conflicts 31

3.6. Poisson Models, Incidence Ratios, Temporal Clustering of Interventions, Models 1–12 34

3.7. Poisson Models, Incidence Ratios, Temporal Clustering of Interventions, Models 13–22 37

3.8. Selected Linear Models, Temporal Clustering of Interventions, Models 1a–9a 39

3.9. ARIMA Models, Temporal Clustering of Interventions 41

3.10. Models with Interaction Terms 42

Summary

Intervention Timing and Temporal Dependence

When Department of Defense (DoD) force planners use integrated security constructs and multiservice force deployment scenarios to estimate force requirements, their models assume that military deployments by U.S. forces occur as independent events that are not systematically correlated over time (i.e., that they are serially uncorrelated). However, the U.S. experiences in Southeast Asia during the 1960s, the Balkans during the 1990s, and the Middle East since 2001 suggest, at least anecdotally, that military interventions may not be serially independent. It is possible, instead, that they demonstrate *temporal dependence*, in which the likelihood of an event in one period is affected by the frequency of similar events in past periods. Temporal dependence would tend to produce event clusters as each incidence of a given event increases the likelihood of similar events in the near future.

If U.S. military interventions do in fact demonstrate temporal dependence, one would expect them to occur in clusters. This would result in a very different "demand signal" for U.S. military capabilities than the current DoD approaches to force planning, which assume serially uncorrelated interventions, produce. This study assesses whether there is empirical evidence of temporal dependence in U.S. military interventions.

Testing for Temporal Dependence

This report tests for temporal dependence by analyzing a set of 66 cases of U.S. Army involvement in "interventions" from 1949 to 2010. For the purposes of this analysis, an *intervention* is an Army deployment of company size or larger for contingency and peacekeeping (or peace enforcement) operations, thus excluding air strikes, airlift, humanitarian operations, sea- and air-based noncombatant evacuations, and small deployments of U.S. military advisers and trainers. To ensure that any finding of temporal dependence is not the result of underlying political (presidential popularity), economic (U.S. gross domestic product and unemployment), or other strategic factors (relative U.S. power, level of conflict, and period), the analysis includes other control variables that may affect the likelihood of military interventions over time. Most important among these is the control for time, specifically the collapse of the United Socialist Soviet Republics (USSR), which marked a significant shift in the overarching geopolitical regime.[1] The analysis also specifies several models of international conflict, using controls for geopolitical regime, global economic health, population growth, and characteristics of the international security context (distribution of power, number of democracies) to assess the existence of dependent clustering in conflict onset that might affect the likelihood of interventions.[2] This analysis used both Poisson and ordinary least squares regressions and yearly data with the count of new interventions (or instances of conflict) as the dependent variable in the analyses.[3]

Results

The results show reasonable evidence that military interventions do occur in temporally dependent clusters but also make it clear that the strength of this dependent clustering effect is heavily influenced by the underlying characteristics and political dynamics of the governing geopolitical regime. The size and strength of the clustering effect varies across specifications, but overall, the empirical results suggest that an additional intervention in one period increases the likelihood of an additional intervention in the next by at least 20 to 25 percent (with an upper bound of as much as 50 percent). This is a fairly significant effect, especially compared with the serially independent distribution of interventions that force planners currently use.

---

[1] The term *geopolitical regime* is used here to denote the fundamental political dynamics of the international system, including the character and distribution of political and economic power, the nature of relationships among major powers, and the security issues around which the system is organized. The Cold War and post–Cold War geopolitical regimes are the two major regimes considered in this analysis.

[2] I use Uppsala Conflict Data Program at the Peace Research Institute Oslo's armed conflict database for the models of political conflict and the control for level of political conflict in the intervention model. The use of a two-stage model is unnecessary because the likelihood of interventions is not strongly associated with the level of instability and conflict.

[3] Regression analysis is a statistical technique used to estimate the relationship between a dependent variable and potential predictors or independent variables, holding other factors constant. The ordinary least squares regression is appropriate when the dependent variable is normally distributed and estimates relationships between variables using a linear approximation. Poisson models are more appropriate for event count data and allow for unequal variance across a distribution.

Autoregressive integrated moving average (ARIMA) models are used with time-series data sets and are designed to capture correlations over time. The ARIMA model estimates the size of temporal correlation in the dependent variable using the correlation of the regression's residuals, or unexplained variation.

Implications for Force Planning

These findings have three broad implications for DoD force planning and the defense analytical community. First, the fact that U.S. military interventions tend to cluster together in time implies that force planning frameworks that assume a serially uncorrelated pattern of interventions, such as the integrated security constructs, may understate actual requirements for forces and capabilities during a period of clustered interventions. This would create operational and strategic risk, and DoD should consider modifying the integrated security constructs to incorporate serial correlation of interventions.[4]

Second, it appears that the likelihood and clustering of interventions are sensitive to the character of a geopolitical regime. This suggests that the pattern of future military operations projected by a force planning framework reflects important, if implicit, assumptions about the nature of the future geopolitical regime. It would be prudent for DoD to make these assumptions explicit and consider whether the existing set of force planning frameworks reflects the spectrum of potential future geopolitical regimes. As a corollary, alternative assumptions about the future geopolitical regime may provide a strategically coherent basis for developing alternative integrated security constructs.

Finally, in addition to quantitative metrics that provide insight into the likelihood of intervention clusters at given time, temporal dependence can be included qualitatively in force planning processes, by incorporating in planning documents an extended discussion of the concept and its implication; how it relates to intervention duration, frequency, and concurrency; and a description of the role the geopolitical regime plays in the likelihood and strength of clustering.

---

[4] Estimates in this report should be considered preliminary and in need of refinement before being used in real force planning analyses.

Acknowledgements

The author thanks Adam Grissom, Lauri Rohn, Michael McNerney, and Steve Watts for their guidance and comments on earlier drafts. Nora Bensahel, Forrest Morgan, and Angela O'Mahony provided extremely insightful comments in their reviews that significantly improved the quality of the report. The author also thanks Dan McCaffrey for his valuable statistical consultation. Finally, Molly Dunigan, Chad Serena, Derek Eaton, Caroline Baxter, Jeff Decker, Stephan Seabrook, and Dan Madden all provided feedback that contributed to the development of this document.

Abbreviations

AR	autoregressive
ARIMA	autoregressive integrated moving average
CINC	Composite Indicator of National Capability
DoD	Department of Defense
GDP	gross domestic product
OLS	ordinary least squares
PRIO	Peace Research Institute Oslo
UCDP	Uppsala Conflict Data Program
USSR	Union of Soviet Socialist Republics

CHAPTER ONE

Introduction

Department of Defense (DoD) force planners use integrated security constructs and multiservice force deployment scenarios to project the numbers and types of demands likely to be placed on U.S. forces in future years. The documents define possible "states of the world," each of which includes steady-state activities and small- and larger-scale contingency scenarios that would require surges in U.S. military forces in a given region. Although these projections do make use of data, models, and simulations, they also rely heavily on two key assumptions—one used to estimate the frequency of future contingencies and the other used to estimate the likelihood the United States will deploy forces overseas. To define the likelihood of different types of scenarios, the current DoD planning process assumes their frequency in the past is the best predictor of their frequency in the future. This approach provides an empirical basis for selecting scenarios but ignores the fact that the nature of conflict may change rapidly over time and may exhibit regional or temporal spillover. To define the likelihood that the United States will intervene militarily, DoD planners consider the level of threat to U.S. strategic interests and the risk associated with the intervention. In this approach, the timing of U.S. military deployments does not follow any systematic pattern or underlying distribution, and each intervention is considered to be largely independent from others.

This does not mean that force planning processes ignore questions about intervention timing. First, they address concurrency or conflict overlap by defining resource constraints. Second, they incorporate duration or conflict length as they map out the phases of a deployment or a contingency. The existing force planning process does not consider, however, whether there is a correlation between military deployments over time that makes the likelihood of future interventions a function not only of U.S. interests and the potential risk but also of the frequency of interventions in the recent past. This relationship would be different from the standard notion of concurrency, in which several deployments occur at that same time, because it would not require overlap between interventions and would be predictable and systematic, not simply a random or chance event.

The question of how the term *intervention* is defined in force planning and in this report is also important. *Interventions* may include many different types and levels of military activity, ranging from the movement of an aircraft carrier from one area to another, to an air strike, to a substantial deployment of ground troops. For the purpose of the empirical analysis in this report, I define *intervention* as a deployment of ground troops, of at least company size. This definition is appropriate because I am most interested in activities that significantly affect the demands on ground troops and that have force planning implications. However, in force planning exercises, interventions may also include smaller, short-term deployments of troops to conduct evacuation operations, the insertion of teams of elite soldiers to accomplish important strategic goals, tactical airstrikes, or other types of operations. The general discussion of interventions and intervention timing in this report refers to these types of military activity as well.

There are empirical, theoretical, and anecdotal reasons to expect relationships between interventions that affect their timing, likelihood, or frequency. First, there are clear examples of events related to military deployments that are not serially independent and that do occur in dependent clusters or waves. For example, several empirical studies of conflict and most empirical work on unconventional threats—such as terrorist attacks—find that the likelihood of future crises, conflicts, and attacks rises when similar events have occurred in immediately preceding years. This type of relationship, in which the likelihood of an event in the present and the future is directly dependent on its incidence in the past, is referred to as *temporal dependence*. Temporal dependence can contribute to event clusters, or uneven, clumpy distributions of events. Temporal dependence observed in terrorist attacks, civil war, and other types of conflict may contribute to similar patterns in military interventions and deployments. There is also evidence of temporal dependence between financial market crises over time, resulting in the wavelike patterns often cited by economists (Corsetti, Pericoli, and Sbraicia, 2005; Caramzaa, Ricci, and Salgado, 2000; Bae, Karolyi, and Stulz, 2003). Parallels between the international economic and political systems make patterns in financial markets relevant to questions about deployment timing.

There is also more direct, qualitative evidence that military interventions and deployments occur in dependent clusters or groups over time. For example, the 1960s saw a string of U.S. military interventions in Southeast Asia, and the early 1990s brought a significant number of U.S. military activities in Bosnia, Haiti, and Somalia. As another example, since 2001, U.S. forces have been involved in numerous small interventions in regionally disparate locations and against terrorist and insurgent groups with anti-U.S. agendas. There are several possible explanations for why such dependent clusters form. Interventions may react to a single set of underlying political factors (the fall of the Union of Soviet Socialist Republics [USSR]) or political instability, or they may reflect an integrated set of policy responses to a single problem (September 11, 2001). Alternatively, the clusters may be driven by the dynamics of the interventions themselves. For example, each U.S. intervention may result in additional instability that demands additional interventions in the near term. Interventions may also trigger changes in the domestic political climate that increase the likelihood or ease of second and third deployments after the first. Finally, interventions may form dependent clusters when one intervention requires supporting interventions to ensure its success.

These examples and possible mechanisms do not prove the existence of temporal dependence or provide information that might guide military planners. However, they do provide a significantly strong enough challenge to the assumption that military interventions are serially independent to warrant additional investigation. The questions of whether and how strongly temporal dependence affects the timing of military interventions should be of interest to military planners because the failure to incorporate this relationship could result in projected force requirements that are too small or do not include the right types of people to meet the demands placed on military personnel. This gap, which emerges when force plans do not account for the rapid increases in demands clustered interventions create, could significantly undermine military readiness and performance.

In this report, I test for the existence of temporal dependence between military interventions and provide some sense of the size of this relationship, defining interventions to include U.S. Army peacekeeping and contingency deployments between 1949 and 2010 (above company size). The next chapter discusses the academic literature on the drivers and timing of U.S. military interventions, the predictors of stability and conflict at the international level, and the economic literature on temporal contagion of financial crises that also informs an understanding of temporal dependence. The third chapter tests for temporal dependence. It describes the data, empirical approach, and results of the empirical analysis, reporting significant evidence of temporal dependence between instances of conflict and instability and military interventions. The fourth chapter discusses the implications of these findings for force planning, including how temporal dependence will affect projected military requirements and how it can be incorporated into planning processes. The final chapter concludes with a discussion of next steps for a research agenda that will further explore the nature and extent of temporal dependence.

CHAPTER TWO

Defining Temporal Dependence: A Review of Existing Evidence

What Is Temporal Dependence?

The argument that military interventions exhibit temporal dependence suggests that these interventions are not independent events but are instead related in systematic and predictable ways over time. More specifically, temporal dependence predicts that the likelihood of an intervention in one year is a function of interventions in previous years. If this relationship is positive, the likelihood of a military deployment increases with the number of interventions in past periods, resulting in clusters of interventions over time. These clusters are not chance events but rather reflect a systematic relationship between interventions in one period and in the next.[1] The interventions in each group may overlap, but temporal dependence does not explicitly require or even predict this overlap, only a change in the likelihood of interventions. The size of an intervention cluster and its density, or the temporal proximity of related interventions, will depend on strength of the temporal dependence. Very strong temporal dependence may have a snowball-like effect, in which the related events aggregate at an increasingly fast rate. Weaker temporal dependence, however, may echo forward only one or two periods before petering out. Temporal dependence is also probabilistic: New interventions always increase the potential for an intervention cluster, but clusters will not always occur and will not aggregate infinitely. The likelihood or probability that the dependent cluster does occur is another measure of the strength of the temporal dependence. Strong temporal dependence may nearly guarantee the existence of a cluster, while weak temporal dependence may result in a cluster only sometimes.

---

[1] Throughout this report, when I refer to clusters, I am referring to clusters caused by temporal dependence, rather than any clusters that may occur by chance or randomly. I oppose this to interventions that are serially independent or that do not have a systematic relationship or correlation over time.

What Does the Literature Say About Intervention Timing and Temporal Dependence?

Existing scholarship on the drivers of U.S. military interventions consider timing and conflict duration when exploring where and when the United States is most likely to intervene but pays limited attention to the potential for or implications of temporal dependence. Literature on political instability, which is relevant to temporal dependence of interventions because instability and conflict create the conditions that demand these interventions, identifies a large number of factors that may affect the likelihood of political conflict and crisis. This literature offers only a limited treatment of regional and temporal spillover that might be associated with temporal dependence. Temporal dependence receives more significant discussion from other disciplines, including economics, and findings from other disciplines can often be generalized to interventions and conflict as well. This chapter discusses each of these major bodies of literature, points to the lack of work focused on temporal clustering of military interventions, and offers some initial explanation for why this gap may be problematic for an understanding of U.S. military interventions.

Interventions and Timing

A review of literature on U.S. military interventions is complicated by the many different definitions of intervention that are used by empirical and theoretical work. While some work describes any military activity as an intervention, others use the deployment of ground forces or include ground forces and air strikes. As already noted, this report focuses only on the deployment of ground forces of at least company size, for which force planning implications are likely to be meaningful. The use of many different definitions likely explains much of the disagreement over the primary factors that explain when and why the U.S. intervenes in some conflicts and not others. Despite the differences in definition, however, existing work on U.S. military interventions suggests three observations that are relevant to the questions of temporal dependence. First, this literature highlights and identifies a large number of international and domestic factors that may contribute to the U.S. decision to use military force. For example, several studies, particularly those focusing on Cold War–era interventions, suggest that international strategic factors, such as arms races or threats to U.S. international interests, along with patterns of political instability are most likely to drive intervention and uses of military force (James and O'Neal, 1991; Brands, 1988). Related studies support these findings, showing that U.S. military interventions are most likely in regions and countries that have strategic resources and areas where the United States seeks to expand its political or economic influence (Yoon, 1997; Pearson and Baumann, 1977; Klare, 1981). Competing arguments suggest, however, that domestic factors, such as upcoming elections, presidential discretion, economic conditions, and domestic attitudes, have the strongest effect on the likelihood of U.S. military intervention (Ostrom and Job, 1986; Fearon, 1994; Meernik, 1994). For example, work focused on presidential discretion argues that the president's personality and the priorities of different administrations play a large role in determining when and where interventions occur (Meernik, 1994). Arguments about the relationship between interventions, economic conditions, and election proximity take several forms (Yoon, 1997; Meernik, 1994). Some studies argue that leaders use interventions to divert attention from weak economic conditions or other problems at home, especially close to elections. Others suggest that economic prosperity encourages military activity but that leaders are hesitant to enter into risky military operations overseas too close to an election. The lack of consensus on these points suggests that the specific context also plays a significant role in intervention time. Finally, there is evidence that the level of domestic support and the existence of an international coalition also increase the likelihood of an intervention (Regan, 1998; Regan and Stam, 2000; Kanter and Brooks, 1994).

Other empirical work on U.S. military interventions suggests that the decision to intervene can be considered as a cost-benefit analysis in which a country is most likely to intervene when it expects the military action to have a low cost and a high probability of success. Factors that affect this calculation include the type of conflict and target, the presence of a coalition or international support, the likely duration of the conflict, and expectations about success (Sullivan and Koch, 2009; Regan, 1998). In general, the United States is more likely to become militarily involved in humanitarian crises than contingencies, in conflicts when there are coalition partners, and when the local violence is not too intense (Regan, 1998). The same cost-benefit analysis also affects the timing of intervention, with interventions being most likely when the intervening party assesses the chances of success to be greatest. Finally, there is evidence that ethnic and religious affinities may also increase the likelihood of intervention, although this factor is less relevant for the United States than for other countries with more homogenous ethnic identities (Kauffman, 1996). Importantly, some of the language used in this body of work is similar to that used in force planning documents.

The second key observation from existing work on military interventions is that military interventions do appear to affect the development, length, course, and intensity of the conflicts in which they occur. Specifically, this work finds that, although diplomatic intervention sometimes reduces the duration of civil wars or is able to truncate periods of instability, military and economic intervention can lengthen conflict duration by providing resources without which one side would likely give up or be defeated (Regan and Aydin, 2006; Bercovitch, 1997; Zartman, 2000; Sullivan, 2009). This does not mean that military intervention is always counterproductive. Intervention may extend a conflict but reduce the level of violence or protect a key strategic interest. However, it does suggest that interventions may have a self-feeding and dynamic quality. By extending the conflicts in which they get involved, they contribute to a demand for continued deployment. A possible inference from this finding is that a given intervention feeds not only itself but also the demand for future interventions. Existing literature does not explicitly address this issue, however.

The final observation from a review of existing work on military interventions is the limited attention paid to timing more generally and the lack of research focused specifically and directly on the question of temporal dependence. While literature on the drivers of U.S. military intervention offers insight into where and why the United States is likely to deploy military force and into the types of domestic and international changes that may affect deployment decisions, this research never explicitly asks whether military interventions are correlated over time or how recent and ongoing interventions affect the likelihood of future ones. Similarly, force size, deployability, and other logistics are not built in as constraints on intervention timing or likelihood. These omissions are significant, especially for studies that include post–Cold War interventions, when U.S. forces are often involved in multiple, smaller, and simultaneous interventions. They also limit the utility of this literature for force planners, who are primarily concerned with questions of resourcing and readiness.

Predictors of Armed Conflict and Political Instability

Understanding the causes and distribution of political instability and conflict is relevant to a study of temporal dependence because these events create demands for military interventions. Even if the United States does not respond to every international conflict (or if it intervenes where no conflict has started), the timing and drivers of instability and crisis events will play roles in the timing of military interventions.

Literature on political instability defines characteristics of the domestic context and the international system that increase the likelihood and duration of instability or conflict and uses forecasting models to predict where and when conflict is most likely. These analyses rely on several rich data sets that record information on conflict start, duration, participants, casualties, and outcomes. Some of these data sets include the Uppsala Conflict Data Program (UCDP) at the Peace Research Institute Oslo (PRIO) database of armed conflict, the Militarized Interstate Dispute data set, and data collected by the Political Instability Task Force. Empirical work on political instability using these data sets produces a myriad of theories and findings on what contributes to instability and conflict. Most of these analyses and their findings focus on a single type of conflict or context. Here, I will summarize some of the key drivers of instability, conflict, and crises across contexts, focusing less on differences across conflict type than on the factors that might contribute to temporal dependence of instability and interventions.

Empirical work identifies political, economic, demographic, and geopolitical factors that may drive or affect the incidence of international conflict, crises, and instability. First, some studies focus on economic determinants of political instability and find that weak domestic economic conditions, as measured by gross domestic product

(GDP) or low levels of economic development, economic protectionism that limits free trade, and domestic income inequality, are associated with a higher probability of such events as coups, riots, protests, and intrastate conflict (Goldstone et al., 2010; Alesina and Perotti, 1993). Other studies suggest that dependence on natural resources and the presence of exploitable commodities (e.g., oil, diamonds) contribute to inter- and intrastate conflict onset and duration and also make states more likely to experience instability caused by violent nonstate groups (Collier and Hoeffler, 2002; Fearon and Latin, 2003; Sambanis, 2005). Studies that focus on political determinants of conflict suggest that conflict and instability of all kinds are more likely to involve or occur in authoritarian countries than in democracies (Goldstone et al., 2010). In addition, strong democracies are typically less likely to enter into conflicts with each other than with authoritarian states and are less likely than than mixed regimes to enter intrastate disputes (Bueno de Mesquita et al., 2003). Incidence of inter- and intrastate conflict, as well as internal violence driven by nonstate groups, is also associated with weak state governance or institutions (Goldstone et al., 2010; Huntington, 1968). Finally, demographic factors also affect the likelihood of conflict and instability. Specifically, the likelihoods of civil war, domestic conflict, and internal instability appear to rise with ethnic diversity and with population size (Goldstone et al., 2010; Collier and Hoeffler, 2005; Sambanis, 2001).

There are also characteristics of the international system that are associated with conflict and instability. For example, a multipolar interstate balance of power (when many states are equally powerful) and world population growth have both been shown to increase the likelihood of instability and conflict (inter- and intrastate) systemwide (Goldstone et al, 2010; Siverson and Sullivan, 1983). Dyadic studies (focused on pairs of countries) of interstate conflict also suggest that power distribution and contiguity are important predictors of interstate dispute. Specifically, conflict appears especially likely between countries that share borders and compete over resources and between pairs of countries experiencing shifts in relative power (i.e., a formerly weak state rising past a formerly strong one) (Siverson and Sullivan, 1983).

Literature that explores the drivers of political instability and conflict considers regional spillover. As noted above, regional spillover can contribute to geographic concentrations of conflict in specific regions. A number of studies have documented evidence of regional contagion and spatial effects that make conflict more likely in countries with neighbors that are also experiencing conflict or instability (Buhaug and Gleditsch, 2008; Ward and Gleditsch, 2000). There are two primary explanations for observed regional spillover. First, instability or intrastate conflict in one country may produce or encourage instability in nearby nations, either because it leads to "copycat" insurrections or because it produces refugees or economic strains that induce conflict in neighboring countries (Buhaug and Gleditsch, 2008). An alternative explanation suggests that certain regions are simply more prone to conflict because of economic, political, or geographic features that they all share. For example, the fact that conflict appears particularly likely in the Middle East may be due to historical religious griev-ances or the presence of lucrative commodities that raise the risk of conflict region-wide. Similarly, the fact that conflict appears especially likely in sub-Saharan Africa may reflect weak governance and widespread poverty throughout the region. The two explanations of regional spillover are not mutually exclusive and might work together to contribute to contagion of conflict. The explanations do have somewhat different implications. While the first suggests a real contagion of political conflict, the second simply emphasizes the importance of regional characteristics in predicting political instability (Buhaug and Gleditsch, 2008; Ward and Gleditsch, 2000).

Attention to temporal dependence between instances of conflict and crisis has been more limited. There are no empirical studies of temporal dependence between instances of political instability or conflicts at the international or global level. How-ever, several studies do identify a wavelike pattern, with periods of general stability punctuated by concentrated periods of overlapping conflicts (Mansfield, 1988; Pollins, 1996). These waves may be created by temporal dependence, but they may also result from other factors, such as youth bulges, economic conditions, or political instabil-ity (Urdal, 2006; Goldstone, 2002). Existing work does not attempt to separate out these different effects. Evidence of temporal dependence at the domestic level is more straightforward. For example, studies that use time-series cross-sectional data on con-flict find significant temporal dependence in the likelihood of civil war within a single country (Beck, Katz, and Tucker, 1998). This type of within-country temporal depen-dence is different from the systemwide temporal dependence that is the focus of this report. However, the existence of temporal dependence between conflicts within states at least justifies a test for temporal dependence between similar events at the interna-tional level.

Literature on political instability provides a good taxonomy for the factors that affect the likelihood of political instability, crisis, and conflict and identifies factors that may indirectly contribute to the timing of military interventions because they are strong drivers of conflict and political crisis. However, its treatment of time and, particularly, temporal dependence remains insufficient. Specifically, existing work does not define the underlying mechanisms that might contribute to the observed temporal dependence between conflicts at the state level, provide good estimates of the size and strength of these temporal and regional spillover effects, or consider the implications of these interdependencies at the international level.

Temporal Dependence in Financial Markets

Literature focused on the spread of financial crises provides some additional insight into temporal dependence, despite its focus on a different type of instability from the one addressed in this report. This literature is especially useful given the dearth of in-depth attention to temporal dependence in studies of conflict. Studies of financial markets and cycles find evidence of waves in which markets across sectors and across the international system move in the same direction, particularly when considering market collapse and financial crises. While some portion of these observed waves can be explained by economic interdependence that extends past national borders, several studies find evidence of temporal dependence that exists even when this interdependence is controlled, suggesting a more complicated relationship (Corsetti, Pericoli, Sbraicia, 2005; Caramazaa, Ricci, and Salgado, 2000). Other studies go a step further and argue not only that contagion exists but also that the degree or extent of contagion is predictable based on underlying characteristics of the economic system (Bae, Karolyi, and Stulz, 2003).

Economic literature also defines several mechanisms to explain the temporal dependence of financial crises that may be relevant to questions asked in this report. This work identifies the structure of economic institutions, political leadership, and individual behavior and expectations as key drivers of temporal contagion in the financial sector (Allen and Gale, 2000; Kodres and Pritsker, 2002; Huang and Xu, 2000). Some studies are even able to quantify the contributions of specific mechanisms to performance. Applied to international conflict and military interventions, this literature suggests that temporal dependence between interventions may be simultaneously driven by many different mechanisms but may ultimately be explained by some combination of domestic and international political, economic, and strategic characteristics. The literature also suggests that it should be possible to place some quantitative bounds, if not on the mechanisms themselves, at least on the size of any temporal correlation.

Summary

Existing work on military interventions and political instability identifies a large number of political, economic, social, and demographic characteristics that contribute to each type of event but pays considerably less attention to either timing or temporal dependence. Political and military leaders and analysts, as well as scholars of conflict, can use existing literature to identify the contexts in which instability or military interventions are likely. However, existing literature does not provide military planners with good or guiding information on the potential existence, size, or implications of temporal correlations between military interventions. Instead, like force planning processes, existing scholarship relies on the untested assumption that these events are independent and can be modeled and considered without direct attention to past interventions. The discussion above challenges this assumption and suggests instead that there are many reasons to expect that not only military interventions but also political instability and conflict are likely to exhibit temporal correlation that leads to clustering in both types of events. The next chapter of this report will offer some empirical assessment of these hypotheses.

CHAPTER THREE

Testing for Temporal Dependence

Methodology

Tests for temporal dependence between military deployments look for relationships between deployments over time. If temporal dependence exists, the likelihood of an intervention in any one period should depend on the number of interventions in previous periods, even once underlying instability, conflict, and relevant political and economic factors are controlled. In this chapter, I test for temporal dependence using a set of 66 U.S. Army "interventions" over the period 1949 to 2010 and a carefully chosen set of control variables.

Although it can be defined in many ways, for the purpose of this report, I define *interventions* to include company-size or larger Army deployments for contingency and peacekeeping (or peace-enforcement) operations. I exclude air strikes, airlift, humanitarian operations, noncombatant evacuation and repatriation operations that involve primarily sea or air forces, and small deployments of U.S. military personnel to advise or train another country's military personnel. These activities are often politically motivated and may even show clustering patterns of their own. However, I have chosen to exclude them here because what I am most interested in are clustered demands that place significant strain on U.S. Army ground troops and that are most relevant to force planners as a result. I set the force size threshold at the company level to exclude deployments that are so small as to place minimal demands on military forces. Even when they occur in clusters, these deployments may not be substantively meaningful. In some cases, data on the numbers of troops involved in small-scale military deployments are not readily available. In these instances, I use secondary sources, other databases, and military almanacs to make the best possible assessment of whether the operation should be included. One goal of future work will be to refine these estimates and, if necessary, the set of interventions included. I used the independent review of my database by two colleagues to validate my coding. It is worth noting that the results are largely robust to small changes in the set of interventions included and are consistent across specifications.

To test the hypotheses outlined in the previous chapter, I conducted two sets of regressions. The first set tested for the temporal dependence of political conflict at the international level, combining instances of inter- and intrastate conflict, and attempted to identify other important predictors of these events. The results of these regressions are important because the underlying distribution and drivers of conflict may also influence the timing of interventions that respond to them. The second set of regressions tested for the temporal dependence of interventions, controlling for the domestic political and economic factors, the international geopolitical factors, and the level of conflict. This approach allowed me to assess the existence of temporal dependence and also offered some insight into the drivers of any temporal relationship that exists. I considered using a formal two-stage model but found this to be unnecessary because, as will be described below, the likelihood of U.S. interventions does not appear to be strongly associated with either ongoing or new conflicts in any given year. Instead, U.S. interventions appear to be more closely associated with characteristics of the overarching geopolitical regime, as well as domestic economic and political factors.

I used both ordinary least squares (OLS) regressions and Poisson models to test for temporal dependence of interventions and conflict. The key dependent variable is a yearly event count, of new conflicts or new interventions, depending on the model.[1] Each intervention or instance of conflict was counted only once, the year in which it began. For example, an intervention that began in 1995 was counted as one new intervention in that specific year (yearly data) but not in subsequent years. Although using data at the monthly level (the level at which it was initially collected) might provide a more nuanced and sensitive test for temporal relationships, yearly models are useful because they aggregate and smooth data in a way that makes evidence of temporal dependence and dependent clustering easy to observe and understand. The nature of deployments makes it difficult to observe and effectively measure temporal dependence at the monthly level. The activation for a deployment may begin well before the deployment start date, and deactivation may also occur over an extended period, making it difficult to precisely capture the timing of an intervention when using monthly data. The other advantage of the yearly models is the breadth of control variables that I could test in the analysis, many of which are not collected at the monthly level. For these reasons, the monthly models are less useful in presenting a complete picture of how the frequency of past interventions affects the likelihood of new interventions and intervention clusters. For the sake of completeness, I did test monthly intervention models. I found these results to be substantively consistent with the yearly models but also more

---

[1] Regression analysis is a statistical technique used to estimate the relationship between a dependent variable and potential predictors, or independent variables. Regressions are valuable for multivariate analysis because they estimate relationships between only two variables, holding other factors constant. The OLS regression is appropriate when the dependent variable is normally distributed and provides estimates for the size and significance of relationships between variables using a linear approximation. Poisson regressions are most often used to model event counts. Rather than assuming equal variance across the entire distribution, as is the case for a linear regression, the Poisson model assumes that the data follow a Poisson distribution and that the variance is equal to the mean of the distribution.

sensitive to changes in specification. Next steps in this research effort will be to rein-vestigate how monthly data might be used more effectively to produce a more sensitive picture of temporal dependence.[2]

Both OLS regressions and Poisson models are used to assess the strength of any dependent clustering and the other political or economic factors that may explain or drive this clustering. While linear models can be used to model event count data, this approach assumes equal variance over the entire period. Imposing a linear model on a distribution with unequal variance could produce results that suggest the existence of dependent clustering where none exists. Essentially, this spurious clustering would reflect not temporal dependence between interventions but the effects of unequal vari-ance in the distribution of interventions over time. The presentation of the results dis-cusses both types of models, differences between them, and why Poisson models are likely more appropriate in both cases. Interpretation of Poisson models can be more complicated than linear ones. I interpreted the results as incidence ratios, which report how much more the estimated incidence rate or how much more likely an "event" will be following a one unit change in any independent variable.

In addition to the standard yearly models that capture temporal correlation with a lagged dependence variable, I also employed autoregressive integrated moving average (ARIMA) models to test for autocorrelation of the residuals, which is another mea-sure of temporal dependence. ARIMA models are used specifically with time-series data sets and are designed to capture correlations over time. If there is a relationship between the specified dependent variable in a given period and the previous period, the ARIMA model will estimate the size of this correlation using the correlation of the regression's residuals, or unexplained variation. If models are appropriately specified, the results derived from the ARIMA models and the yearly regressions with lagged terms should be largely similar.[3]

Data and Operationalization

I specified separate models for interventions and instability. This section describes the data and controls included in the analysis.

Interventions

The intervention data set includes 66 interventions between 1949 and 2010. It com-bines interventions reported in several different existing data sets (including Regan,

---

[2]   Monthly regression results are available on request.

[3]   I used the ARIMA function in Stata for this piece of the analysis. The function fits a model in which distur-bances are allowed to follow a linear autoregressive specification. I fit an AR1 model, which means I included only one autoregressive lag (denoted AR in result tables) but retained the original dependent variable and include no moving average component.

1998, and Sullivan and Koch, 2009) with those cited in Congressional Research Service reports that summarize War Powers Resolution reports made by the President since the 1970s (Grimmett, 2002; Grimmett, 2010). To ensure that my set of interventions was comprehensive, I consulted additional reports produced by the Center for Defense Information (Berry et al., 2001; Marte and Wheeler, 2007). To compile my final data set of U.S. military interventions, I merged and reconciled these different lists, addressed discrepancies, and recorded conflict start and end dates and conflict type, distinguishing between contingency operations (including counterinsurgency) and peacekeeping interventions. Figure 3.1 shows the number of military interventions each year included in the data set. Each intervention is counted only in its year of onset. Several observations are worth noting. First, the frequency and number of interventions is greatest in the period between 1988 and 2004. There are several explanations for this trend, including the end of the Cold War and the fall of the Soviet Union, which made the risk of intervention somewhat lower; the related rise of the United States as the world's major superpower; and, starting after 2001, U.S. operations against al-Qaeda and other terrorist organizations, in Iraq, Afghanistan, and elsewhere. Lower rates of intervention over the 1960s and 1970s may reflect the drain on resources created by the Vietnam War and a period of détente with the Soviet Union and rapprochement with China, relationships that significantly suppressed the desire of all three countries for involvement in proxy wars. A second important observation is the drop in new interventions after 2004. This likely reflects the effects of the significant demands on U.S. military personnel and resources created by the wars in Iraq and Afghanistan. Finally, presidential discretion may be another important consideration.

Figure 3.1
Interventions, by Year of Onset

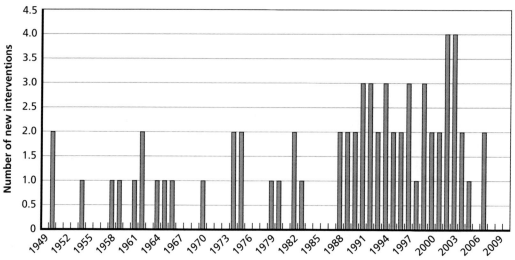

RAND *RR192-3.1*

Presidents George W. Bush and William J. Clinton are often viewed as having been more "intervention minded" than others.

As noted above, the independent variable in all specifications is the number of new interventions in a given year. To measure the existence and strength of temporal dependence, I used the number of interventions in the prior year, called the *lag*, as an explanatory variable. I considered one-, two-, and three-period lags (the number of interventions one, two, and three years prior) to assess the duration of any temporal correlation.

Previous work on the timing of military interventions suggests a number of other control variables that can affect when interventions occur and that may drive or explain apparent clustering. For instance, the collapse of the Soviet Union was one event that significantly altered the likelihood of interventions because, as suggested above, it affected the international distribution of power, the challenges and constraints the United States faced, and the geopolitical regime. Figure 3.1 confirms that the number of interventions per year does increase markedly starting in 1988, several years after Gorbachev became head of the Communist Party (1985) and after the Geneva (1985), Reykjavik (1986), and Washington (1987) summits signaled a substantial thaw in superpower relations.[4] Looking at the empirical data, Figure 3.1 suggests that the clearest shift in U.S. intervention behavior occurred between 1987 and 1988.[5]

In addition to a control for the Cold War, there are several other political and economic variables that past work has used to explain intervention timing. First, I considered two economic controls, annual change in U.S. unemployment and annual U.S. GDP growth. These variables address the competing arguments in existing literature on the effect of the economy on the likelihood of intervention. In both cases, I included the lagged form of the variable because, when the decision to intervene is made, decisionmakers can act only on available information, in this case the previous year's economic data (and possibly projections about the current year). Second, to capture the effect of presidential popularity (and associated political discretion), I included the lagged change in the president's average annual approval rating, essentially whether the president's approval rating was trending upward or downward in the previous year. I chose this measure over other possible measures of presidential popularity because it is one that presidents may be especially sensitive to and one that can be compared

---

[4]   However, inserting a Cold War control into the statistical model of intervention timing is complicated by the fact that there is disagreement over when the Cold War actually "ended." Although the United States and the Soviet Union did not formally recognize the end of the Cold War until the Malta Summit in December 1989, the geopolitical regime had clearly already begun to shift before that meeting.

[5]   To be conservative and because, as will be shown in detail later, the results are somewhat sensitive to the way this control is specified, I considered specifications using 1987, 1988, and 1989 as the "end" of the Cold War. This approach assumes that the geopolitical regime was shifting during this period. I also examined other possible end dates, for example 1985, 1986, 1990, and 1991. These results did not vary significantly, however, from the 1987 (1985, 1986) and 1989 (1990, 1991) results that I include, so I will not discuss them directly.

across different presidential administrations more easily. These controls address arguments about the risks and rewards of military deployments for political figures facing reelection.

I included the lagged annual change in the U.S. Composite Indicator of National Capability (CINC) score, a measure of U.S. power or capabilities from the Correlates of War database, as a potential predictor of military interventions because both the international distribution of power and the position of the United States in the international system are likely to affect willingness to intervene.[6] The CINC score is a relatively crude measure of U.S. power and one that trends down over time. However, I chose to use this measure to be consistent with previous work and to establish a baseline that could be refined in further research. One goal for future work might be to develop a better measure of U.S. capabilities. I used the lagged change in CINC score to partially account for the fact that the CINC score trends downward over time, making the raw index score a proxy for time rather than a measure of the level or change in U.S. power. Finally, I included a control for the overall level of international conflict using the lagged cumulative number of ongoing conflicts in each year according to the UCDP-PRIO Armed Conflict data set, which includes all instances of armed conflict, inter- and intrastate, between 1946 and 2010. This data set is described in more detail below.

Almost as important as what I have included as controls is what I have intentionally excluded. First, unlike other models of intervention timing, I do not have separate controls for ongoing interventions or specific controls for Vietnam, Korea, or the Gulf War, which some past work has argued affected the likelihood of new interventions. I did not include these variables because they are too closely associated with the most important dependent variable, the lagged intervention term. Because I was most interested in assessing the existence of temporal clustering, I could not also include additional controls for these specific conflicts. Second, I did not control for specific events that a qualitative reading might suggest influenced decisionmaking about interventions. This choice prevented overspecification of the model, which could have led to spurious or misleading results, especially since the data set itself is already relatively small.

In part to deal with some of these same concerns about overspecification and additional concerns about multicollinearity, my approach to specifying intervention models was intentionally parsimonious. I began with simple specifications that considered only the effect of the lagged dependent variable and then compared more-complex models to these simple ones to understand clearly how the geopolitical regime, global conflict, and domestic political and economic conditions affect any observed temporal

---

[6] The *National Material Capabilities Data Documentation* codebook provides details on how the CINC is constructed. Most basically, it includes six indicators: military expenditure, military personnel, energy consumption, iron and steel production, urban population, and total population. See Correlates of War Project, 2010, and Singer, Bremer, and Stuckey, 1972.

dependence. The close relationship between certain controls and the end of the Cold War, specifically conflict and the CINC score, is another issue that I considered and tried to disentangle in the analysis and results.

Armed Conflict

Models of political conflict use the UCDP-PRIO's most recent armed conflict data set, which was updated in April 2012. This data set includes all instances of conflict, including inter- and intrastate. From 1946 to 2011, a total of 212 new armed conflicts were initiated, which are defined as "a contested incompatibility that concerns government and/or territory where the use of armed force between two parties, of which at least one is the government of a state, results in at least 25 battle-related deaths" (Uppsala University, Department of Peace and Conflict Resolution, 2011, p. 1; see also Gleditsch et al., 2002). I used these data to define two variables: a count of new conflicts started in each year and a count of total ongoing conflicts. As above, each conflict was counted only once, in the year it began. I combined intrastate and interstate war for the analyses presented here, despite the fact that most empirical work considers one or the other of these two, because what I was most interested in is whether there is any temporal correlation in conflict at an international, systemic level, irrespective of whether that conflict is inter- or intrastate.

Figure 3.2 shows the number of new armed conflicts initiated over the period covered by the intervention data set, while Figure 3.3 shows the pattern in total ongoing conflicts. There is some evidence of a wavelike pattern that would be consistent with a clustering hypothesis, but significant variation from year to year weakens the

Figure 3.2
Armed Conflict, Onset

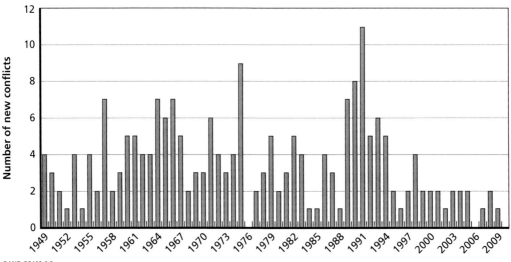

Figure 3.3
Armed Conflict, Cumulative

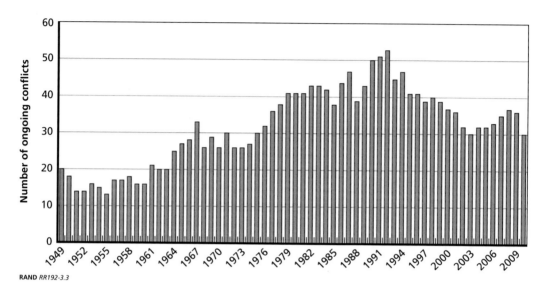

clarity of this pattern and makes it difficult to locate specific peaks, valleys, or clusters. Some general observations are still possible. First, conflict appears more likely in the 1960s, early 1970s, and 1990s than in other decades. Second, as in Figure 3.1, the number of conflicts rises significantly around the end of the Cold War, starting in 1989 after a short lull in the mid-1980s. This may be important when considering appropriate control variables and the cutoff date for any Cold War control, as discussed above.

An interesting empirical question may be whether the increase in conflict incidence occurred before, after, or at the same time as the change in U.S. intervention behavior. If the increase in conflict incidence occurred before the change in interventions, the increase in U.S. intervention rate may reflect a U.S. response to a change in the behavior of other states following the weakening of the Soviet Union. If the increase in conflicts occurred after the increase in U.S. interventions, it may be the change in U.S. behavior that is driving the rise in conflict incidence, either because interventions encourage conflict or because changes in U.S. behavior signal to smaller states a shift in the geopolitical regime. If the number of conflicts and interventions occurred together, both may be most strongly affected by external geopolitical changes.

Figure 3.3 shows a gradual increase in the number of ongoing conflicts from 1949 to 1991, with small dips in the mid-1970s, late 1980s, and mid-2000s. There is a longer, more gradual decline until 2004, when the total number of ongoing conflicts appears to increase once again. The wavelike pattern of conflict is easier to see in this ongoing conflict graph, but it is still difficult to determine whether or not there is temporal dependence in political conflict.

The specification of the armed conflict models is similar to that for the intervention models. The tests for temporal dependence between interventions use the number of new armed conflicts in each year as the key predictor and the lagged number of armed conflicts in the prior year as the primary predictor. Relevant controls in this case will be somewhat different, but will still include a control for changes in the geopolitical regime, or the end of the Cold War. Figure 3.2 shows that, for conflicts, the clear shift in state behavior occurs in 1989, with a sharp increase in the number of new conflicts, following a lull in 1988. However, because the distribution of armed conflict appears to vary not only across the Cold War divide but also across decades, it is also necessary to explore the use of several decade controls. For example, as noted above, conflict appears more likely in the 1960s and 1990s but somewhat less likely in the 2000s. To explore this observation empirically, I also used specifications that control for each individual decade.

I used the key findings of past work on armed conflict, including the literature described in Chapter Two, to guide my selection of other relevant controls, again taking a parsimonious approach, limiting the inclusion of variables that may lead to overspecification and paying attention to concerns about multicollinearity between the Cold War control and other predictors that may be closely associated with this shift in geopolitical regime or with time more generally. The discussion in Chapter Two identified a number of relevant factors, including economic conditions, commodity dependence, population growth, regime type, and distribution of power. Many of these variables will be closely associated with the end of the Cold War or with a given period, including population growth, the CINC score, and the number of democracies, forcing caution in their inclusion and interpretation. While the first two variables fell consistently over the period under consideration, the number of democracies rose rapidly after 1988. I addressed this by comparing alternate specifications and by using first-differenced variables that consider the rate of change over time rather than the absolute level of a given variable. To measure distribution of power, I included the change in U.S. capabilities relative to other major powers (the CINC score). I also included a measure of world population growth (percentage growth over the past year).[7] I used the Polity IV database to develop a raw count of number of democracies in the international system in each year under consideration. Finally, I used two measures of global economic health. The first measures change in global GDP (taken from the International Monetary Fund's World Economic Outlook Database), and the second measures the change in total trade exports (from the United Nations' International Trade and Merchandise Statistics Data). These variables can be used to assess arguments about the effects of interdependence and economic prosperity on conflict.

---

[7] Because the dependent variable is total instances of conflict or instability at an international level, using world population growth is more appropriate than using country-specific measures of population size, growth, or density. If I were trying to explain variation in conflict rates across nations, these country-specific indicators would be the more appropriate control variables.

Each of the above variables, population growth, global economic changes, and the number of democracies, are included as lagged variables using the same logic as in the intervention models. Specifically, conflicts in a given year react to economic and geo-political characteristics that prevailed over previous periods.

Finally, I experimented with including a control for the lagged number of U.S. military interventions. Including this variable along with the lagged number of conflicts raises some concerns, since it may essentially control twice for new instances of conflict that also trigger interventions. I tested these specifications to provide some insight into possible relationships and interpreted the results with caution.

As for the intervention models, I took a parsimonious approach to specification, starting with simple models that included only the dependent variable and its lag, then added the control for the Cold War, and then built more complex models with all relevant variables, addressing multicollinearity concerns along the way.

Results

This section describes the results of the empirical analysis, presented in Tables 3.1 through 3.10. A few notes on interpretation of the regression tables will facilitate a reader's understanding of the results.[8] In most cases, the primary indicator of temporal dependence is the lagged dependent variable, listed first in each regression table. It reports the size and strength of the relationship between instances of instability or military inventions in a given period and the period immediately preceding it. For the linear models, interpretation is relatively straightforward. A one-unit increase in any dependent variable is associated with a change in the independent variable equal to the regression coefficient. For example, in an intervention model, if the coefficient on the lagged intervention variable is "1," one additional intervention in one period is associated with an increase of one intervention in the next. Temporal dependence exists when this term is statistically significant. In the ARIMA specifications, also included in the presentation of the results, the autoregressive (AR) component listed at the bottom of the table identifies the size and significance of any temporal dependence. The interpretation of the AR term is the same as just described for the lagged term. If the coefficient on the AR component is "1," one additional intervention in one period is associated with one additional intervention in the next.

Interpretation of the other variables included in the tables is similar. The coefficients report the size of the relationship between that control and the dependent vari-

---

8   The tables report regression coefficients, with standard errors in parentheses. The asterisks are used to indicate statistical significance based on the t-statistic, with more asterisks suggesting a greater level of confidence in the results. Following convention, one asterisk suggests significance at the 0.1 level (90 percent confidence), two asterisks the 0.05 level (95 percent confidence), and three the 0.01 level (99 percent confidence). The caret indicates the 0.15 level (85 percent confidence).

Table 3.1
Poisson Models, Incidence Ratios, Temporal Clustering of Conflicts, Models 1–5.1

	Model					
	1	2	3	4	5	5.1
Lagged conflict	1.1** (0.04)	1.1** (0.04)	1.1** (0.04)	1.1*** (0.04)	1.04 (0.04)	1.04 (0.04)
Cold War 1987		1.2 (0.2)				
Cold War 1988			1.13 (0.2)			
Cold War 1989				1.24 (0.23)		
1960s					1.56** (0.35)	1.37* (0.23)
1970s					1.31 (0.35)	
1980s					1.07 (0.29)	
1990s					1.47 (0.4)	1.3 (0.3)
2000s					0.50*** (0.12)	0.48*** (0.1)
Wald χ^2	7.1	7.6	7.2	9.6	44.8	44.4
N	61	61	61	61	61	61

NOTE: Standard errors in parentheses; * $p < 0.10$, ** $p < 0.05$, *** $p < 0.01$.

able, and the asterisks report statistical significance. For variables that can take only the values "0" and "1" (which is the case for most of the period and event controls), the regression coefficient reports the change in the dependent variable associated with the change in the control from "0" or "not true" to "1" or "true."

Interpretation of the Poisson regressions will be somewhat different. In this case, I have interpreted the results as incidence ratios, so the results in the tables report for each variable the likely change in the estimated rate of incidence of the dependent variable, either a new intervention or new armed conflict, following a one-unit increase in the dependent variable. For example, if the coefficient on the lagged intervention term in a Poisson model is 1.2, one additional intervention in one year increases the expected incidence of intervention in the next period by a factor of 1.2. This means that the risk of additional conflicts increases by about 20 percent over the base rate with an additional intervention in the first period. For other variables, the interpretation is again the same. The coefficient on the Cold War variable, for example, will report how much more or less likely conflicts are after the Cold War ends than while it is ongoing. If this coefficient were 0.5, the expected incidence of interventions in the

Table 3.2
Poisson Models, Incidence Ratios, Temporal Clustering of Conflicts, Models 6–16

	Model										
	6	7	8	9	10	11	12	13	14	15	16
Lagged conflict	1.1** (0.04)	1.01 (0.04)	1.1* (0.14)	1.03** (0.04)	1.1*** (0.04)	1.03 (0.04)	1.1*** (0.04)	1.04 (0.04)	1.1*** (0.04)	1.03 (0.04)	1.05 (0.04)
1960s		1.5** (0.26)		1.18 (0.25)		1.5** (0.27)		1.4** (0.24)		1.36* (0.23)	
1990s		3.5*** (1.3)		1.66* (0.5)		1.33 (0.33)		1.23*** (0.3)		1.01 (0.33)	
2000s		1.78 (0.9)		0.65 (0.23)		0.46*** (.09)		0.42 (1)		0.35*** (0.1)	
Lagged number of democracies	0.99** (0.002)	0.98 (0.1)									0.99* (0.01)
Lagged change, global exports					1.01 (0.01)	1.01* (0.01)					1.01* (0.01)
Lagged change, global GDP					1.0 (0.06)	0.95 (0.05)					
Lagged change, CINC							1.75 (2.7)	0.31 (0.52)			0.22 (0.39)
Lagged world population growth rate			1.9*** (0.45)	1.97 (0.96)							1.8^ (0.7)
Lagged U.S. interventions									1.02 (0.06)	1.16^ (0.12)	1.26*** (0.1)
Wald χ^2	14.8	59.1	25.9	46.4	11.5	49.6	7.8	44.6	7.13	62.03	38.9
N	61	61	61	61	61	61	61	61	61	61	61

NOTE: Standard errors in parentheses; * $p < 0.10$, ** $p < 0.05$, *** $p < 0.01$, ^ $p < 0.15$.

Cold War period would be 0.5 times that of the post–Cold War period, or about one-half as likely, holding all else constant.

What Drives Armed Conflict?

The results of the models of armed conflict show little evidence of temporal dependence between instances of armed conflict once the effects of variation in period and the geopolitical regime are properly controlled. Across specifications, the most important predictors of conflict incidence are controls for specific decades with high and lower levels of conflict, suggesting that conflict may occur in wavelike patterns based on geopolitical context rather than small, contained clusters. Such variables as the number of democracies, relative U.S. power, economic growth and trade flows, and U.S. military interventions are relevant to conflict timing but have weaker explanatory power and offer somewhat less insight into clustering. The relevance of these variables to patterns of conflict may be best understood through the filter of period or as a set of geopolitical factors that characterize specific periods that explain why conflict is more likely at certain points than others.

The simplest Poisson model in Table 3.1, model 1, shows the temporal dependence in armed conflict considering only the number of new conflicts, with no controls. The incidence ratio of 1.1 is statistically significant but substantively small (the likelihood of a conflict increases only 10 percent). A comparison between this model and the same model using a linear regression, shown in model 1a in Table 3.3, illustrates a significant difference between the two. The simple linear model shows a coefficient of 0.37 on the lagged term, suggesting that an addition of three conflicts (above the mean or expected value) in one year would lead to one additional conflict in the next. The difference between the two specifications suggests that variances are not equal across the armed conflict distribution and that the clustering picked up by linear models may be due primarily to this unequal variance. Since the Poisson model is likely more appropriate for the event count data, I will focus on Poisson models going forward but will briefly cover linear specifications toward the end of the section.

The next set of specifications, models 2, 3, and 4 in Table 3.1, compares the results of the three different Cold War controls, using 1987, 1988, and 1989 as the "end years," respectively. The results are largely similar to each other. In each case, the inclusion of the period control has almost no effect on the lagged intervention term and is not statistically significant. This would suggest that the change in the geopolitical regime at the end of the Cold War had little effect on the rate of conflict overall or on any clustering of conflict that occurred. However, a closer look at the decade-specific controls reveals that geopolitical regime and time period are, in fact, important to explaining the timing and distribution of conflict. Specifically, controlling for each decade separately (model 5, Table 3.1) confirms that conflict is significantly less likely in the 2000s than during any other period (by a factor of 0.5) and more likely in the 1960s (by factor of 1.6). The 1990s control also has an incidence rate greater than one (1.47), suggesting a higher rate of conflict, although it falls short of statistical significance. The controls for the 1970s and 1980s are further from statistical significance, and the 1950s control is omitted as the referent group. Including these controls reduces the size and eliminates the significance of the lagged conflict term. These results suggest that the clustering observed in the initial specifications may be driven by the existence of decades in which conflict is significantly more or less likely than at other points. These periods of high and low conflict likelihood may reflect shifts in the geopolitical regime (as may be the case in the 1990s) or by specific events (such as 9/11 and subsequent U.S. military operations). Because of these results, I used the decade controls in the remainder of the conflict models but include only the two controls that were significant (2000s and 1960s) and the one that was on the borderline of statistical significance (1990s). This choice reduces concerns about overspecification, especially in models that include additional covariates. Model 5.1 in Table 3.1 confirms the similarity of the results using only these three decade controls.

Tests of other controls confirm the relevance of the decade controls to the timing of conflict and explanations of any observed clustering and offer insight into other possible determinants of conflict timing and likelihood. Diagnostic tests suggested that the democracy variable, CINC score, and change in world population are highly correlated with the decade controls. As a result, including all variables in the same model may cause multicollinearity problems and misleading results. To address this concern, I considered control variables both on their own and as part of more-complex models.

The count of democracies and world population growth (both included using a one-period lag) are statistically significant when they are the only variables in the model but lose this significance once decade controls are added. The independent effect of number of democracies (model 6, Table 3.2) is negative but substantively small, while that of population growth is somewhat larger, with an incidence ratio of 1.9 (model 8). This suggests that a one-percentage-point increase in population growth rate will increase the incidence of conflict in the next year by a factor of 1.9. Neither variable affects the size or significance of the lagged conflict term, suggesting that neither can explain clustering. In both cases, however, once decade controls are added, the significance of numbers of democracies and world population growth disappears (models 6 and 7, Table 3.2). Also interesting is the fact that the inclusion of the democracy and population controls affects the size and significance of the decade controls. This is indicative of the close correlation between decade and both number of democracies and population growth. To the extent that population growth and number of democracies affect the rate of conflict, they matter through the filter of time.

Next, I consider the two economic variables, lagged change in global GDP and lagged change in global trade flows. Concerns about correlation between change in global GDP and change in trade are lower than for population and democracy variables. This is especially true of changes in trade flows. A model that includes the decade controls and both economic indicators, model 11 in Table 3.2, finds that the lagged change in exports is positive and significant but that its effect is substantively small. The change in GDP variable is not significant, but its direction suggests that as global GDP rises, the incidence of conflict falls. The inclusion of the economic controls also shrinks the size and significance of the lagged conflict term, implying that overall global economic prosperity may be another factor, in addition to the geopolitical regime, that explains the small amount of clustering observed in the first specification. Neither economic control is significant on its own in models without decade controls (model 9, Table 3.2). This suggests that even the effects of economic variables may be best interpreted as functions of period, in this case, decade.

The final relevant indicators are those associated with U.S. power and influence, the lagged annual change in CINC score and new interventions. I used the change in CINC index to address the fact that this variable declines slowly and consistently over time and so serves as a proxy for year, rather than the relative level of U.S. strength. The change in CINC score variable is never significant, regardless of whether it is included alone or with the period controls (models 12 and 13, Table 3.2). The control for U.S. interventions is worth interpreting with some caution since it may be that there is a close association between interventions and instances of armed conflict, as well as one between interventions and time. Although this control is not significant included on its own or without decade controls (models 14 and 15, Table 3.2), it approaches significance in the model that includes decade controls and is in the positive direction, implying that an increased number of U.S. military interventions in one period may be associated with additional risk of conflicts in the next. Once again, the empirical results are clearly sensitive to specification. This sensitivity appears to reflect strong correlations between predictor variables and the importance of the overarching geopolitical circumstances as an important predictor of conflict.

The final Poisson specification includes all tested controls except the decade controls. Since several controls—democracy, world population growth, CINC score—are correlated with time, including all together may provide a partial replacement for the decade controls themselves and provide additional insight into how these covariates independently affect conflict timing. Model 16 in Table 3.2 suggests several interesting results. First, the lagged intervention term is positive and significant. An increase of one additional intervention in the past period increases the estimated incidence rate of conflicts in the next by about 1.26. Both the number of democracies and the percentage change in trade exports are also statistically significant, but the effects of both are substantively small. The world population growth term and CINC score fall short of statistical significance. Finally, the inclusion of this set of variables eliminates the significance and reduces the size of the lagged conflict term and, therefore, the clustering effect. To the extent that dependent clusters of conflict exist, they can be explained by underlying characteristics, such as the number and types of states, economic circumstances, and geopolitical considerations, including U.S. interventions in this case. These characteristics vary over time, specifically across decades for instances of armed conflict.

Testing for Robustness: Linear and ARIMA Specifications

To explore the robustness of these results, I compared them to selected linear specifications. Tables 3.3 and 3.4 report the results from the linear model versions of selected Poisson specifications. The substantive implications and patterns of significance are similar. Models 2a, 3a, and 4a test the three versions of the Cold War control. None is significant, and evidence of dependent clusters of conflict remains. As noted already, the size of the clustering coefficient in the linear specifications implies a larger effect than the Poisson models. In this case, each additional conflict in the past increases the expected number of conflicts in the future by about 0.37. In substantive terms, an increase of about three conflicts in one period would be likely to increase conflicts in the next period by one. However, also as noted above, the linear models may be

Table 3.3
Linear Models, Temporal Clustering of Conflicts, Models 1a–5.1a

	Model					
	1a	2a	3a	4a	5a	5.1a
Lagged conflict	0.37** (0.16)	0.36** (0.04)	0.37** (0.17)	0.37** (0.17)	0.14 (0.18)	0.16 (0.17)
Cold War 1987		0.49 (0.6)				
Cold War 1988			0.3 (0.6)			
Cold War 1989				0.64 (0.59)		
1960s					1.6** (0.83)	1.24* (0.72)
1970s					0.9 (0.97)	
1980s					0.21 (0.89)	
1990s					1.4 (1.1)	1.0 (0.99)
2000s					−1.33* (0.72)	−1.68*** (0.5)
Constant	2.09*** (0.52)	1.83*** (0.57)	1.92*** (0.6)	1.7*** (0.51)	2.4*** (0.84)	2.8*** (0.62)
F-statistic	5.5	3.3	3	4.38	8.76	13.57
N	61	61	61	61	61	61

NOTE: Standard errors in parentheses; * $p < 0.10$, ** $p < 0.05$, *** $p < 0.01$.

Table 3.4
Linear Models, Temporal Clustering of Conflicts, Models 6a–14a

	Model								
	6a	7a	8a	9a	10a	11a	12a	13a	14a
Lagged conflict	0.32* (0.17)	0.06 (0.04)	0.28* (0.18)	0.11 (0.17)	0.38** (0.15)	0.13 (0.17)	0.37** (0.16)	0.12 (0.18)	0.2 (0.19)
1960s		1.6** (0.71)		0.75 (0.84)		1.7** (0.72)		1.23* (0.73)	
1990s		5.4*** (2)		1.9 (1.2)		1.02 (1.1)		0.35 (1.3)	
2000s		4.29* (2.4)		-0.31 (1.04)		-1.63*** (0.5)		-2.25*** (0.73)	
Lagged number of democracies	-0.02** (0.008)	-0.09*** (0.03)							-0.02 (0.02)
Lagged change, global exports					0.04 (0.04)	0.05 (0.4)			0.02 (0.03)
Lagged change, global GDP					-0.02 (0.24)	-0.21 (0.24)			
Lagged change, CINC									0.22 (0.39)
Lagged world population growth rate			2.1*** (0.8)	2.5^ (1.7)					2.0 (1.5)
Lagged U.S. interventions							0.1 (0.2)	0.4 (0.29)	0.63*** (0.23)
Constant	3.5*** (0.9)	7.2*** (1.7)	-0.97 (0.99)	-1.4 (1.23)	1.8** (0.8)	3.1*** (0.99)	2.0 (0.6)	2.7 (0.62)	-0.34*** (3)
F-statistic	7.7	13.35	8.1	10.4	3.61	44.6	7.13	62.03	7.6
N	61	61	61	61	61	61	61	61	61

NOTE: Standard errors in parentheses; * $p < 0.10$, ** $p < 0.05$, *** $p < 0.01$, ^ $p < 0.15$.

unequal variances across the distribution of armed conflict rather than the existence of true dependent clusters.

While none of the Cold War controls is significant, the decade controls are, with conflict being more likely in the 1960s and 1990s (although only the coefficient on the 1960s term reaches statistical significance) and less likely in the 2000s. The inclusion of decade controls (model 5a, Table 3.3) eliminates the significance and reduces the size of the lagged conflict term. Once again, this suggests that the most significant predictor of conflict timing and a powerful explanation for why conflicts appear to cluster is period: Conflict is more likely in certain decades than others. As in the Poisson models, a specification that omits the insignificant controls (model 5.1a, Table 3.3) for the 1970s and 1980s is largely the same as that controls for each decade. To avoid overspecification, I included controls only for the 1960s, 1990s, and 2000s in subsequent specifications.

Including the control for the number of democracies in model 6a (Table 3.4) had a small negative effect on the rate of conflict but almost no effect on the clustering term. The size of the democracy term's effect appears somewhat larger in the linear specification when period controls are included but has essentially no substantive significance on its own. This suggests that, as in the Poisson models, the effect of regime type or the number of democracies on the timing of conflict is best understood through the filter of time.

The effect of the population growth rate is also similar in Poisson and linear specifications. While this rate has a large, positive statistically significant effect on incidence of conflict when included on its own, it is somewhat less significant in models that also use period controls. Model 8a (Table 3.4) shows that, when this variable is the only one in the model, a 1-percentage-point change in population growth rate is associated with an increase of about 2.1 additional conflicts in the next period. The lagged number of conflicts retains significance in this specification. However, the inclusion of relevant period controls in model 9a (Table 3.4) absorbs the significance of both terms. Like the number of democracies, the relationship between population growth and international conflict may reflect its strong correlation with time and the variation in level of conflict across decades.

The economic terms also behave in linear as in Poisson models, although both fall short of statistical significance in this case, and they do not affect the lagged conflict term, the measure of dependent clusters in these models (models 11a and 12a, Table 3.4). Finally, neither the change in CINC index nor the lagged number of interventions is significant when included on its own or in models with the decade controls (models 12a and 13a, Table 3.4).

The final linear specification (model 14a, Table 3.4) mirrors the final Poisson specification (model 16, Table 3.2), by including each of the tested covariates but none of the period controls. In this case, the only significant predictor is the lagged U.S. interventions term. Each of the other covariates falls short of statistical significance, and the lagged conflict term also loses its significance. Comparing this result with model 5a (Table 3.3), which only used decade controls, shows that the size of the coefficient on the lagged intervention term is largely the same. This suggests that variables strongly correlated with time, such as the change in CINC score, population, and number of democracies, may partially replace decade controls. However, models with only decade controls are generally a better overall fit of the data, suggesting that they capture the underlying characteristics of specific geopolitical periods better than more disaggregated measures.

Finally, Table 3.5 presents selected ARIMA specifications. It confirms the results already described and is nearly identical to linear specifications. As a reminder, in these

Table 3.5
ARIMA Models, Temporal Clustering of Conflicts

	Model						
	1b	2b	3b	4b	5b	6b	7b
Cold War 1987		0.77 (0.8)					
Cold War 1988			0.2 (1.1)				
Cold War 1989				0.75 (0.92)			
1960s					1.44** (0.67)		1.34** (0.64)
1990s					1.2 (1.1)		5.3** (2.4)
2000s					−1.93*** (0.52)		4.44^ (2.8)
Lagged number of democracies						−0.02 (0.02)	−0.09*** (0.04)
Lagged change, global exports						0.02 (0.03)	0.03 (0.03)
Lagged change, global GDP							
Lagged change, CINC							
Lagged world population growth rate						2.2 (1.5)	1.7^ (1.21)
Lagged U.S. interventions						0.67*** (0.24)	0.4^ (0.28)
Constant	3.4*** (0.42)	2.9*** (0.65)	3.3*** (0.88)	2.9*** (0.74)	3.33*** (0.44)	0.13 (3.44)	3.5^ (2.5)
AR (lagged conflict)	0.37** (0.16)	0.36** (0.16)	0.36** (0.17)	0.35** (0.17)	0.16 (0.16)	0.2 (0.17)	0 (0.14)
σ	2.1*** (0.22)	2.09*** (0.23)	2.1*** (0.22)	2.09*** (0.23)	1.94*** (0.22)	1.95*** (0.22)	1.77*** (0.17)
N	62	62	62	62	62	61	61

NOTE: Standard errors in parentheses; * $p < 0.10$, ** $p < 0.05$, *** $p < 0.01$, ^ $p < 0.15$.

models, the size and strength of any clustering effect are revealed by the AR term, which measures the correlation between the residuals, called autocorrelation. Models 1b through 5b (Table 3.5) show that, as in each previous specification, there is evidence of dependent clusters when only Cold War controls are included and in models that omit any control for time. The inclusion of the decade controls, however, eliminates evidence of systematic clustering. In model 5b (Table 3.5), the lagged AR term is not significant, but the results show that conflict is more likely in the 1960s and less likely in 2000s. Models 6b and 7b (Table 3.5) consider the effects of including other controls, both with and without the decade variables. Again the results are the same as above. There is no evidence of clustering once number of democracies, world population growth, change in global exports, and U.S. interventions have been controlled, and the most consistent and significant predictors of conflict timing are the controls for period. Models 6b and 7b (Table 3.5) also weakly suggest that the number of U.S. interventions may increase the likelihood of conflict and that the number of democracies may have a small negative effect.

Summary

There is very little evidence that new instances of armed conflict cluster in regular or systematic patterns, over time, once changes in underlying geopolitical conditions and global economic conditions are controlled. Although there is some evidence of statistically significant clustering when tested without any controls, the most important predictors of the incidence of conflict are the decade controls, suggesting that it is the combined effect of the political and economic characteristics of the geopolitical regime that are most likely to affect the rate of conflict. These results hold across specifications. While there is some sensitivity to the selection of variables included, the relevance of the period to observed clustering and the observation that the effects of other variables can be explained through the filter of period or geopolitical regime appear fairly robust.

As noted in Chapter One, U.S. military planners generally assume that the likelihood and nature of future conflict can be predicted using the frequency of conflict in the past. The results suggest that this is a reasonable assumption only if planners are sure that the "past" on which they base their analysis has the same fundamental geopolitical characteristics as the one they are currently operating within. Any shift in geopolitical regime may fundamentally alter the timing, likelihood, distribution, and nature of conflict. It may be difficult for planners and decisionmakers to diagnose changes in regime immediately—they may only be observable with the benefits of hindsight. The results suggest that regime type, distribution of power, and U.S. military behavior may partially affect the incidence and rate of conflict. Tracking these characteristics over time may allow planners and decisionmakers to quickly diagnose regime shifts that may affect the rate of conflict. In addition, planners and decisionmakers may be able to increase their "early warning" capabilities by building a deeper or more nuanced understanding of the defining characteristics of the current geopolitical regime.

Is There Temporal Dependence Between Military Deployments?

The results suggest stronger evidence of temporal dependence in military interventions than in armed conflict but also reveal that the results are sensitive to characteristics of the underlying geopolitical regime. Apart from the lagged intervention term, which is significant in most specifications, the strongest predictors of intervention incidence are the controls for period, specifically the control that splits the Cold War period from the post–Cold War period.[9] The inclusion of this Cold War control reduces the size and weakens the significance of the lagged intervention term, suggesting that, when controlling for a higher post–Cold War incidence of military interventions, evidence of temporally dependent clusters is somewhat weaker. This result implies that the formation of dependent clusters is strongly affected and even driven by the overarching geopolitical regime. Models that include an interaction term between the Cold War indicator and the number of interventions offer additional evidence that clustering due to temporal dependence is more likely after the Cold War ends than before.

Other control variables, including those for economic prosperity, underlying conflict, and U.S. capabilities, have much weaker effects that are, as above, best interpreted through the filter of time, or as characteristics of a specific geopolitical regime that explain why interventions are more likely at some points than others. The results suggest further that interventions are more likely after the Cold War than during it and that dependent clusters may characterize certain regimes more than others. If the current geopolitical regime is one characterized by clustering, planners and decisionmakers have incentives to understand why it occurs and how it may affect U.S. national security and demands on military forces.

Model 1 in Table 3.6 shows the simplest Poisson specification in which only the lagged intervention term is included. This model suggests significant evidence of temporally dependent clusters: One additional intervention is associated with a 1.53 increase in estimated incidence rate of interventions in the next or an increase of 53 percent over the base rate. Since the base rate is about 1 intervention per year, this amounts to an increase of about 0.5 interventions in the next year due to temporal dependence. This is largely similar to the effect estimated using the linear model (model 1a, Table 3.8), suggesting that the assumption of equal variance over the distribution may be reasonable in this case. To be conservative and because the Poisson specifications may provide the most accurate estimates of clustering, I will focus on these models in the discussion of the results.

The next three models in Table 3.6, models 2, 3, and 4, show the Cold War controls. As noted above, I consider three possible cutoff points for the Cold War—1987, 1988, and 1989. Although the lagged intervention term remains significant in all cases,

---

[9]   I also used decade controls, but t-tests of the coefficients in these models suggested that the differences between the decades during the Cold War (1950s, 1960s, 1970s, 1980s) were not significantly different from each other. The same is true for the 1990s and 2000s.

Table 3.6
Poisson Models, Incidence Ratios, Temporal Clustering of Interventions, Models 1–12

	Model											
	1	2	3	4	5	6	7	8	9	10	11	12
Lagged Interventions	1.53*** (0.12)	1.25** (0.14)	1.22* (0.14)	1.22** (0.14)	1.49*** (0.12)	1.26** (0.13)	1.24* (0.15)	1.27** (0.15)	1.59*** (0.15)	1.32** (0.15)	1.3** (0.16)	1.27** (0.15)
Cold War 1987		0.41*** (0.12)								0.38** (0.12)		
Cold War 1988			0.41*** (0.13)			0.45** (0.14)	0.46** (0.15)				0.4** (0.13)	
Cold War 1989				0.45*** (0.14)				0.51** (0.17)				0.45** (0.15)
Lagged U.S. GDP growth					0.93 (0.04)	0.97 (0.05)	0.97 (0.05)	0.96 (0.05)				1.01 (0.05)
Lagged unemployment					0.93 (0.08)	0.97 (0.09)	0.98 (0.09)	0.97 (0.08)				
Lagged change in CINC												
Lagged change in presidential approval									1.02** (0.01)	1.02** (0.01)	1.02* (0.01)	1.02* (0.01)
Lagged total conflict												
Wald χ^2	28.78	33.54	31.84	32.84	30.6	35.82	34.44	35.35	30.47	39.87	35.22	36.48
N	61	61	61	62	61	61	61	61	60	60	60	60

NOTE: Standard errors in parentheses; * $p < 0.10$, ** $p < 0.05$, *** $p < 0.01$.

it shrinks significantly once the Cold War controls are included, from an incidence ratio of 1.53 to 1.22 or 1.25, depending on the specification. This suggests that including the control for the geopolitical regime soaks up about half of the observed clustering effect. Including this control also reduces the significance of the lagged intervention term, effectively weakening certainty about the existence of dependent clustering. The results suggest that the likelihood of clustering between interventions is strongly affected by the characteristics of the overarching geopolitical system. The implications of this observation for planners and for the interpretation of other controls will be traced out in more detail below. In addition to their effect on the clustering term, the Cold War controls are also strongly associated with the incidence of interventions. They are also significant in each case and comparable in size, between 0.41 and 0.45. The incidence rate is less than one, suggesting that the estimated incidence of interventions during the Cold War is about one-half that of the post–Cold War period.

In addition to the overarching geopolitical regime, there are a number of other factors that may affect the timing of interventions and the likelihood that dependent clusters form. First, economic controls test the relevance of the diversionary theory of war, or the notion that presidents may use force to divert attention from weak economic conditions. I test two economic controls, lagged change in GDP and lagged change in unemployment. Lagged change in unemployment never approaches statistical significance in any specification. Table 3.6, models 5, 6, 7, and 8, shows the results for models that include U.S. GDP growth. This variable comes closest to significance when it is included alone, and its value suggests that it reduces the likelihood of interventions. This result would be consistent with the diversionary theory of war (war is more likely as the economy does poorly). When included with Cold War controls, however, the change in GDP term does not seem to affect intervention incidence or temporal dependence between interventions. This does not mean that resource constraints are irrelevant, only that governing economic conditions cannot be used to predict the likely timing of interventions or to explain observed clustering.

Controls for political popularity test for the effect of presidential support and approval on the likelihood of military interventions. I use the lagged change in public approval of the president because it is a metric that is likely to be of significant interest to a governing president—a president who has falling public support may be more cautious about the use of force than one whose popularity is increasing. It is also a metric that is easy to compare across administrations and over time. The results in Table 3.6, models 9, 10, 11, and 12, show that the lagged change in presidential approval does appear to increase the likelihood of an intervention, suggesting that, as presidents become more popular, they are more likely to use military force overseas. The size of the effect is somewhat small: A 1-percentage-point increase in popularity increases the estimated incidence for interventions in the subsequent year by only 1.02, or 2 percent over the base rate. This effect remains regardless of the cutoff used for the end of the Cold War. Interestingly, however, it also increases the size of the lagged intervention term, which suggests that, once we control for variation over time in trends in presidential approval and the effect of these trends on intervention timing, the underlying temporal dependence between interventions is actually stronger and more noticeable. This reinforces the importance of presidential discretion and underlying political characteristics in explaining the incidence of military interventions and even driving temporal dependence. However, there are limits to the effects of domestic political factors on the likelihood of military interventions. Controls for election year and the year prior to an election are not significant in any specification.

The change in CINC score never approaches statistical significance and has only a very small substantive effect, regardless of specification (Table 3.7, models 13 through 17). As noted previously, the CINC score declines consistently over time and is a crude measure of U.S. capabilities, so the results should be interpreted with some caution. Its inclusion does not affect the size or significance of the lagged intervention term but does slightly reduce the incidence ratios for each of the Cold War controls. This suggests that the difference in intervention likelihood across the two regimes may be greater when controlling for changes in U.S. relative power and that the CINC score, even if not significant on its own, may matter as a characteristic of the geopolitical regime. Thus, U.S. capabilities may have some effect on the willingness to intervene, but this effect may be most relevant as a characteristic of the geopolitical system.

Finally, new and ongoing armed conflicts may affect the timing of U.S. military interventions, since armed conflicts create demands for new interventions. The onset of new armed conflicts is never statistically significant and does not affect the overall strength of the temporal dependence between interventions, regardless of whether the term is included with the Cold War control or not. As shown in Table 3.7, the lagged number of cumulative conflicts is significant only when included with no Cold War control (models 18 and 19), but its effect on interventions is small (increases likelihood of conflict only about 4 percent) and does not eliminate the significance of the lagged intervention term (although it does slightly reduce its size). Once the Cold War controls are included, it is significant only in the specification using 1989 as the cutoff point, and even then, its significance is fairly weak, and it is still small. This suggests that, while the likelihood of interventions may increase slightly during periods of high conflict, the timing or concentration of conflicts does not explain clustering. As is the case of other potentially important domestic political and economic variables, the level of conflict may matter most as a characteristics of a governing geopolitical regime.

The results suggest that the overarching characteristics of the geopolitical regime itself play a significant role in determining when clustering of interventions occurs and when interventions are less likely to exhibit temporal dependence. The inclusion of an interaction term between the Cold War indicator and the number of interventions in a given year tests the hypothesis that the likelihood of temporally dependent clustering is significantly different in the Cold War and post–Cold War periods. The interaction term is generated by multiplying the Cold War indicator and the number of interven-

Table 3.7
Poisson Models, Incidence Ratios, Temporal Clustering of Interventions, Models 13–22

	Model									
	13	14	15	16	17	18	19	20	21	22
Lagged Interventions	1.57*** (0.13)	1.58** (0.16)	1.3** (0.15)	1.27** (0.16)	1.32** (0.17)	1.42*** (0.13)	1.46** (0.16)	1.33** (0.17)	1.3** (0.16)	1.33** (0.18)
Cold War 1987			0.35*** (0.12)					0.49** (0.18)		
Cold War 1988				0.37*** (0.13)					0.53* (0.21)	
Cold War 1989					0.43*** (0.16)					0.61** (0.23)
Lagged U.S. GDP growth		0.96 (0.04)	1.04 (0.06)	1.03 (0.06)	1.2 (0.06)		0.96 (0.05)	1.0 (0.05)	1.0 (0.05)	1.0 (0.05)
Lagged unemployment										
Lagged change in CINC	1.00 (0.02)	1.02 (0.003)	0.98 (0.03)	0.98 (0.03)	0.99 (0.03)					
Lagged change in presidential approval		1.02* (0.01)	1.02* (0.01)	1.02* (0.01)	1.02* (0.01)		1.02* (0.01)	1.02** (0.01)	1.02* (0.01)	1.02** (0.01)
Lagged total conflict						1.04*** (0.01)	1.04*** (0.01)	1.02 (0.02)	1.03^ (0.02)	1.03* (0.02)
Wald χ^2	31.09	27.75	46.5	39.03	39.5	32.64	41.5	41.17	43.61	45.11
N	60	60	60	60	60	60	60	60	60	60

NOTE: Standard errors in parentheses; * $p < 0.10$, ** $p < 0.05$, *** $p < 0.01$.

tions for each year. The resulting term will take a value of "0" in the post–Cold War period and will equal the number of interventions in the Cold War period. For simplicity, I included only one version of this conditional specification, which uses 1988 as the end of the Cold War. Although alternative end dates and specifications with more variables offer slightly different results, the results for this model are generally representative of the overall significance of the interactive or conditional relationship.

The model that includes the interaction term will also include each of its constituent parts, so the full specification can be expressed:

$$y(interventions) = b_1 \text{ lagged interventions} + b_2 \text{ Cold War indicator}$$
$$+ b_3 \text{ lagged interventions} \times \text{Cold War indicator}.$$

For the post–Cold War period, the effect of lagged interventions (or the amount of temporal dependence) will be just equal to the coefficient on the "lagged interventions" term. For the Cold War period, the amount of temporal dependence will be the joint effect of the coefficients on the "lagged interventions" (b_1) and interaction term, (b_3).

The results of model 23 (Table 3.10) suggest support for the conditional hypothesis and offer evidence that clustering due to temporal dependence is more likely and stronger after the Cold War than during it. The lagged intervention term is positive and significant, suggesting an increase of 1.26 in the incidence ratio. Clustering is likely, therefore, in the post–Cold War period. The interaction term is negative (implying a reduction in the effect of lagged interventions on those in the current period) and falls short of statistical significance. A t-test of whether the sum of the two coefficients (the total effect of interventions during the Cold War) is significantly different from "0" cannot reject the null hypothesis of no effect or no temporal dependence. Thus, while there is evidence of temporally dependent clustering after the Cold War, there is little evidence of this relationship during the Cold War period. However, these results should be interpreted with caution and as suggestive rather than definitive because the sample is relatively small, especially in specifications with the interaction term, which essentially splits the Cold War and post–Cold War period and considers the existence of temporal dependence in each period separately.

Testing for Robustness: Linear and ARIMA Specifications

While the Poisson models are likely the more appropriate specification, a comparison with linear models may still provide a useful robustness test. Across the board, the results are largely similar. Model 1a (Table 3.8) shows the results including only the lagged intervention term. It is statistically significant and relatively sizable. A one-intervention increase in one year is associated with an expected increase of about 0.5 intervention in the next year. In more meaningful terms, two additional interventions in one year are likely to trigger at least one in the next. The inclusion of the Cold War controls in models 2a, 3a, and 4a (Table 3.8) also shows patterns similar to the Pois-

Table 3.8
Selected Linear Models, Temporal Clustering of Interventions, Models 1a–9a

	Model								
	1a	2a	3a	4a	5a	6a	7a	8a	9a
Lagged Interventions	0.52*** (0.11)	0.45** (0.13)	0.31** (0.14)	0.27* (0.15)	0.3** (0.15)	0.35** (0.14)	0.3** (0.15)	0.32** (0.15)	0.46*** (0.14)
Cold War 1987			−0.9*** (0.37)			−0.91** (0.42)			
Cold War 1988				−0.96** (0.38)			−0.88* (0.45)		
Cold War 1989					−0.88** (0.38)			−0.78* (0.45)	
Lagged U.S. GDP growth		−0.03 (0.04)				0.03 (0.05)	0.02 (0.05)	0.02 (0.05)	−0.03 (0.05)
Lagged unemployment									
Lagged change in CINC						−0.02 (0.03)	−0.03 (0.03)	−0.03 (0.03)	−0.0004 (0.02)
Lagged change in presidential approval		0.02** (0.01)				0.02** (0.01)	0.02* (0.01)	0.02* (0.01)	0.02* (0.01)
Lagged total conflict		0.04*** (0.02)				0.01 (0.02)	0.02 (0.02)	0.02^ (0.02)	0.04** (0.01)
Constant	0.51*** (0.15)	0.19 (0.32)	1.3*** (0.36)	1.4*** (0.43)	1.33*** (0.42)	0.75 (0.55)	0.74 (0.6)	0.66 (0.6)	0.19 (0.44)
F-statistic	25.32	13.00	17.65	16.7	15.9	8.95	9.4	9.2	10.22
N	61	60	61	61	61	60	60	60	60

NOTE: Standard errors in parentheses; * $p < 0.10$, ** $p < 0.05$, *** $p < 0.01$, ^ $p < 0.15$.

son models. In each case, the Cold War control is significant and negative: Interventions are less likely during the Cold War than after (shifting between periods decreases the number of expected interventions by about one in each case). The inclusion of the Cold War control decreases the size and significance of the lagged intervention term. In these specifications, an intervention in any year increases the likelihood of an additional intervention in the following year by about 30 percent. The effect is the weakest when the cutoff for the Cold War is set at 1988. However, the control for geopolitical regime does not eliminate the significance of the lagged intervention term or the observed clustering in linear models.

Table 3.8 shows linear specifications that include the other relevant controls. In general, the direction, size, and significance of these other controls are similar to those for the Poisson models. In each case, the lagged intervention term is significant, and its effect size ranges from about 0.3 to 0.35 (one additional intervention in a given year increases interventions in the next by about 0.3, or the risk of intervention rises by about 30 percent), similar to that observed in Poisson models. The Cold War controls are negative, and their size is also comparable or slightly larger: The expected number of interventions is likely to decrease by slightly less than one conflict (compared with 0.6 in Poisson models) with the change from post–Cold War to the Cold War period. Changes in presidential approval have a small positive effect on the likelihood of interventions. Positive GDP growth and larger U.S. CINC scores are not statistically significant when included in models with other variables. Finally, the lagged overall level of conflict is never significant when included with Cold War controls. The final linear model, Model 9a (Table 3.8), considers the effects of other controls when they are included without the Cold War controls. The results do not change much. The major differences are that the lagged total conflict terms become significant but are still small, and the size and significance of the lagged intervention term increases. The results of the linear models, then, largely confirm the results of the Poisson models, both in terms of the size of effects and their overall significance.

Finally, I considered three ARIMA specifications, using the 1988 Cold War end date. I chose this particular end date for the ARIMA robustness tests because it showed the weakest evidence of temporally dependent clusters in linear specifications and because it is the midpoint of the three Cold War controls, which have performed similarly in all previous specifications. The first specification (model 1b, Table 3.9) includes only the lagged intervention term. It shows a result nearly identical to the linear model above. The lagged AR term is significant, and its 0.53 coefficient is similar to previous specifications including only the lagged intervention term. Models 2b and 3b (Table 3.9) include the Cold War control. Model 3b also includes GDP growth (for substantive reasons), presidential approval, and the Cold War controls. Once again, the results are substantively the same and have coefficient sizes and significance nearly identical to those for previous linear models. In both, the Cold War control has a strong negative effect. Also in both, the lagged AR term is positive and significant (although weakly

Table 3.9
ARIMA Models, Temporal Clustering of Interventions

	Model 1b ARIMA	Model 2b ARIMA	Model 3b ARIMA
Cold War 1988		−1.3*** (0.36)	−1.0** (0.42)
Lagged U.S. GDP growth			−0.008 (0.04)
Lagged unemployment			
Lagged change in CINC			
Lagged change in presidential approval			0.01* (0.01)
Lagged total conflict			0.03* (0.1)
Constant	1.03*** (0.25)	1.9*** (0.32)	1.24* (0.65)
ARIMA L.AR	0.53*** (0.1)	0.26* (0.14)	0.33** (0.16)
σ _cons	0.96*** (0.07)	0.9 (0.26)	0.84 (0.07)
N	62	62	60

NOTE: Standard errors in parentheses; * $p < 0.10$, ** $p < 0.05$, *** $p < 0.01$.

when only the Cold War control is included). The effect size is also similar: One additional intervention in one year triggers about 0.3 additional interventions in the next. Change in U.S. GDP has no substantive effect, while change in presidential approval and the total level of conflict have small, weakly significant positive effects. Overall, then, ARIMA specifications confirm the robustness of the results presented earlier. There does appear to be evidence of intervention clustering, although this effect is strongly affected by the underlying geopolitical regime and weakly affected by domestic political considerations.

Finally, model 23a (Table 3.10) offers the linear version of the conditional specification with the interaction term that tests for the existence of a significant difference in the extent of temporal dependence observed in the Cold War and post–Cold War periods. The results once again suggest that the likelihood and strength of temporal dependence are greater in the post–Cold War period. The lagged intervention term is still positive and significant, while the interaction term is negative and just significant. This suggests that the size of any temporal dependence during the Cold War will be significantly smaller than that observed after the Cold War ends. Furthermore, a t-test of their combined effect again cannot reject the null hypothesis (that there is no temporal dependence during the Cold War). As above, therefore, the inclusion of the interaction term provides at least some evidence that clustering due to temporal dependence

Table 3.10
Models with Interaction Terms

	Model 23 Poisson (incidence ratio)	Model 23a Linear
Cold War 1988	0.47** (0.17)	−0.56 (0.47)
Lagged Interventions	1.26** (0.14)	0.42** (0.18)
Lagged interventions × Cold War[a]	0.85 (0.18)	−0.38* (0.23)
Wald χ^2	37.6*** $p < 0.00$	
F-statistic		11.73*** $p < 0.00$
N	61	61
Test: Lagged Interventions + Lagged Interventions × Cold War = 0	$\chi^2 = 0.1$ $p = 0.8$	F = 0.1 $p = 0.73$

NOTE: Standard errors in parentheses; * $p < 0.10$; ** $p < 0.05$, *** $p < 0.01$.

[a] The interaction term is created by multiplying the two constituent parts, the number of military interventions and the time period

is more likely and stronger after the end of the Cold War. However, the results should again be interpreted with some caution due to the relatively small sample.

Summary

The results show reasonable evidence that military interventions do occur in temporally dependent clusters but also make it clear that strength of this clustering is heavily influenced by the underlying characteristics and political dynamics of the governing geopolitical regime. These characteristics define the international political system and affect how and when force is used. The geopolitical regime appears to affect both the incidence of interventions and their distribution, or the likelihood of clustering. First, interventions are less likely during the Cold War than after. The political and strategic dynamics that prevailed during the Cold War explain why this might the case. The Cold War was a bipolar geopolitical regime in which the United States and the USSR acted carefully to retain and expand their own influence without upsetting the other. Both the United States and the USSR were hesitant to intervene and risk provoking the other. Afterward, however, in a new geopolitical regime, the United States took on a different role and was increasingly willing to use force overseas. The way the United States used military force may also have changed at this point, shifting from more traditional deployments of conventional forces to shorter-term, smaller scale, unconventional operations, often in response to violent nonstate actors, terrorist attacks, and failed states emerging in the wake of the Soviet Union's collapse.

The control for the Cold War also affects the size and strength of the temporally dependent clustering observed in military interventions. This suggests the temporally dependent clustering may be more likely in certain types of geopolitical regimes than others and is a relevant concern to planners and strategists only at these times. Models that include interaction terms provide some additional evidence that clustering is more likely and stronger in the post–Cold War period than during the Cold War, although caution is warranted in interpreting these results due to the small sample. If it is the case that the current geopolitical regime is one in which clustering is more likely, today's military planners may need to consider the effects of temporal correlation more carefully than their predecessors.

There is also some evidence that certain domestic political and, potentially, economic factors affect the incidence of interventions and perhaps clustering of interventions as well. For example, the results suggest that presidents are generally more likely to intervene when their political support has been rising than when it has been falling in the recent past. There is a clear political survival logic to this observation. Since most presidents suffer declining popularity over their terms, this may also mean that interventions are more likely early in a president's tenure than later, when they may have less political capital to exploit.

If clustering is a phenomenon that is characteristic of specific geopolitical regimes, the key question for policymakers and defense planners becomes identifying the observable characteristics of regimes, where temporally dependent clustering of interventions is likely, as well as ways to identify when shifts in geopolitical regimes that might change the strength or relevance of clustering have occurred. The results in this report suggest several possible characteristics, including underlying levels of conflict, relative distribution of power, economic prosperity, regime type, and even the nature of deployments (long versus short, traditional ground combat versus surgical strike, etc.). The results also suggest that the primary triggers for geopolitical regime change are events, including state failure or formation or the start or end of ongoing wars that shift the international balance of power or the orientation of major states. However, the results also make it clear that regime change is rarely an overnight event: Many occur over several years and affect state or conflict behavior only with some sort of lag.

CHAPTER FOUR

Implications for Force Planning

Will Temporal Dependence Affect Force Requirements?

The results in the previous chapter suggest that, at times, U.S. military interventions exhibit temporal dependence that increases the chances they will occur in dependent clusters rather than following the serially independent distribution assumed in force planning processes. This result appears somewhat sensitive to the characteristics and political dynamics of the governing geopolitical regime and, as a result, may not always be a relevant concern for force planners and decisionmakers. However, where it does occur, clustering may have significant effects on force requirements that are worth considering in some additional detail.

To understand the implications of temporal dependence on the sufficiency of projected force requirements, a more detailed review of some key force planning assumptions will be useful. As noted in the introduction, current DoD planning processes rely on broad national security and defense guidance, experience, some operational data, and input from military commanders to define specific military requirements for each type of operation and each regional command. Force planning processes consider three types of activities: (1) steady-state operations that occur continuously and set the baseline for forces required in a region, (2) smaller contingences, and (3) larger-scale operations. The sum of these estimates across contingencies and across regions together results in the projected force requirements used to guide force planning decisions. Requirements associated with the latter two types of activities are based on two key variables: (1) the likelihood that these contingencies occur and (2) the propensity or likelihood that the United States will become militarily involved.

Current processes estimate the likelihood of various contingencies by assuming that the distribution of contingencies in the future will look similar to that of the recent past. Since the nature of conflict is constantly evolving, this assumption is almost certainly incorrect. Furthermore, the assumption does not address or incorporate the effects of temporal dependence observed in the empirical analysis in the previous chapter. Importantly, while it was not the focal point of this report, the finding that political instability and conflict, when considered in aggregate terms at the international level, also exhibit temporal dependence is important for force planners.

Even if the timing and frequency of instability and conflict on the international level do not directly cause new interventions, they still affect the context in which the military operates and in which military interventions occur. While current force planning documents assume a distribution of instability and conflict in the future that is similar to that in the past, findings in this report suggest that future conflict and instability may not simply mirror past events but may sometimes build off of them. A failure to account for this spillover may lead to contingency projections that are too small and military force projections that misestimate the numbers and the types of capabilities needed to meet actual demands.

Force planning assumptions about the likelihood that the United States will become militarily involved in a contingency rely primarily on qualitative assessments of U.S. interests, the threat the contingency poses to these interests, and the risk associated with the operation. For example, interventions are more likely when the threat is high and the risk lower than in alternative contexts. While these assumptions are consistent with some of the findings in existing literature, they do not consider temporal dependence or intervention timing in any systematic way. As noted elsewhere in the report, this means that the likelihood and force requirements of a given intervention are independent.[1] This approach ignores the temporal dependence observed in the previous chapter. Specifically, force planning processes that incorporate temporal dependence must address both the direct force demands of a given deployment and the heightened risk of future demands due to temporal spillover. Force planning that does not address both components risks generating force estimates that underestimate the number of interventions, the demands on military personnel, and the force size needed to meet future demands.

The extent and severity of this underestimation will depend on the rate at which deployments group together and the size of the deployments involved in the group. If deployments group or cluster at a rapid pace or if the interventions involved in a specific group are large, failing to account for temporal dependence may significantly underestimate force demands. On the other hand, if deployments overlap and group together more slowly and if the interventions in each group involve few U.S. troops, ignoring the temporal dependence of interventions may have fewer implications. The results in the previous chapter suggest a rate of dependent clustering that is sizeable enough to have noticeable effects on military force requirements, especially when compared to the serially independent distribution of interventions used in current planning processes and especially if overlapping conflicts that do occur place demands on the same occupations. For example, recent deployments in Iraq and Afghanistan drew heavily on special operations, civil affairs, and military police, and the overlap in the two conflicts forced a rapid increase in personnel trained or serving in these areas. Fail-

---

[1] The exception is directly overlapping interventions, "concurrency" in planning documents, in which resource constraints are often factored in.

ure to adjust force planning processes to account for temporal dependence increases the likelihood that similar shortfalls will occur in the future.

Mechanisms of Temporal Dependence

Understanding the mechanisms that may drive temporal dependence are important first steps to incorporating the relationship into force planning. Although the results in this report do not provide complete clarity on which of several possible mechanisms has the most explanatory power, they do provide some insight into factors that may affect the strength of temporally dependent clustering. The results suggest that temporal dependence of interventions has its roots in the characteristics and political dynamics of the governing geopolitical regime: the prevailing international and domestic political, economic, and strategic factors that characterize a geopolitical system. The analyses in the previous chapter also identify a number of more specific drivers of temporally dependent clustering that could be investigated in future work. For example, the results suggest the importance of such factors as the distribution of power among states, underlying levels of conflict and regime change, population growth, attitudes toward the use of force, and economic prosperity as characteristics of a given regime that may affect intervention timing. It is also possible that certain types of interventions are more likely to cluster than others. The apparently different rates of clustering in the post–Cold War period, for instance, may reflect differences in the types of interventions that occur in the pre– and post–Cold War periods. For force planners, the potential sensitivity of clustering to geopolitical regime increases the importance of building a complete understanding of the specific political characteristics and dynamics that make different geopolitical regimes distinct and the identifying signals that mark a shift in regime even more important.

However, simply pointing to the geopolitical regime as the driver of temporally dependent clustering does not explain the more-specific mechanisms that explain why and how temporal dependence between interventions occurs during periods when it exists. First, it is possible that U.S. military interventions, when they occur, contribute directly to additional demands on U.S. forces, either by encouraging additional instability or because military interventions often require secondary, supporting interventions in nearby areas to ensure the success of the primary intervention. There is some empirical evidence for this mechanism in the models of armed conflict presented previously. Specifically, in several specifications, the lagged number of U.S. interventions had a positive effect on the likelihood of new instances of conflict. While this may be explained by the fact that both interventions and conflict are most likely in the 1990s, it may also indicate that the decision to deploy U.S. forces does affect conflict dynamics. The fact that military interventions may inherently create demands for additional interventions does not necessarily mean that military deployments "make things worse." It may be that U.S. intervention behavior is a signal to other states about what they can or cannot get away with. However, the relationship does suggest that the decision to deploy forces may have unexpected and indirect consequences over an extended period.

A second possible mechanism for temporal dependence is an "in for a penny, in for a pound" effect among either the public or policymakers involved in making decisions about the deployment of U.S. forces. According to this story, a perceived (and perhaps real) drop in the marginal costs and consequences of additional military deployments once one intervention occurs lowers political barriers and public resistance to second and third deployments. This lower threshold makes multiple overlapping deployments more likely and may result in the observed temporal dependence. There is also some empirical support for the link between political capital and interventions that indirectly bolsters a mechanism focused on domestic political processes. Specifically, there is evidence that high and rising presidential popularity, a measure of political capital, increases the likelihood of deployments and may contribute to clusters in some cases.

Third, temporal clustering of military interventions may reflect strategic sets of grouped U.S. responses to international security challenges. Specifically, if the United States often responds to international security challenges with multiple and complementary deployments, the result will be groups of interventions with a single goal or purpose. A qualitative review of intervention clusters in the data set suggests some evidence for this explanation of temporal dependence, including the post–9/11 military response, responses to communist movements in the late 1960s early 1970s, and deployments following the collapse of the USSR in the early 1990s. However, a "strategic response" mechanism is unable to explain all observed intervention bundles.

A final mechanism for temporal dependence has to do with the credibility of commitments. Specifically, deployments and interventions may occur in clusters because, once a deployment has occurred, political and military leaders feel pressure to deploy additional forces elsewhere to support similar or related goals. Under this mechanism, interventions occur in temporally correlated groups because, once committed, the United States cannot avoid making additional commitments without significant negative effects on its reputation.

How Can Temporal Dependence Be Integrated into the Planning Process?

The data and empirical results in this report do not allow me to finally disentangle the mechanisms above to identify the primary drivers of observed clustering, beyond stressing the importance of the geopolitical regime to both its existence and its strength. However, it is still possible to make some preliminary observations about how temporal dependence can be incorporated into force planning.

Assessing the Relevance of Temporal Clustering

The results suggest that temporal clustering of military interventions may exist at certain points and not at others. A first important challenge for policymakers and planners, therefore, may be simply to identify the international, geopolitical, and domestic factors that make clustering more likely at certain points rather than others. The results here suggest several potential drivers of temporally dependent clustering, but additional work is needed to solidify and complete this list. First, the distribution of power and the relative power of the United States appear important. Second, the underlying level of conflict and international political dynamics (number of democracies, number of failing states) may also be relevant. Third, U.S. attitudes toward the use of force and its strategic orientation (both difficult to measure in an empirical model) also affect the nature of the geopolitical regime. Fourth, the nature of deployments may matter. Interventions of certain lengths or with certain types of goals may be more likely to cluster than others. Finally, domestic political dynamics, elections, presidential discretion, and presidential personality may also affect the likelihood of interventions and clustering. Future empirical work may develop better proxies for these potential geopolitical drivers of temporally dependent clusters and use these variables to clarify what it is about geopolitical regimes that makes clustering more likely at certain points than others.

Planners may also need to identify when major shifts in geopolitical regime that will have implications for temporal clustering have occurred. Some shifts may be obvious, following major revisions of the balance of power, such as the collapse of a major power or its defeat in a war. It is also possible that an economic collapse, like the Great Depression, or a major event, such as 9/11, that affects attitudes toward the use of force might trigger a shift in geopolitical regime. On the other hand, it may be that shifts in geopolitical regime occur only gradually, driven by the slow decline of an old power or the gradual rise of a new one, an enduring economic or population boom in a concentrated region, or longer-term shifts in geopolitical priorities. These types of shifts might be much harder to detect or observe, even in hindsight. As it relates to temporal dependence, identifying the factors that contribute to changes in the overarching regime will help force planners and policymakers understand when interventions and clustering in interventions are most likely and when temporal dependence will most severely affect the demands on U.S. forces and resources.

Building Temporal Dependence into Force Planning

When clustering is relevant, however, there are some additional steps that force planners might take to incorporate temporal dependence into force planning processes. Temporal dependence may affect many aspects of projected force requirements, including the necessary size of the force, the appropriate capability mix, and even the basing of troops at the international level. Building temporal dependence into the force planning process could reduce the risk of shortfalls in specific occupations or at least provide a tool force planners can use to diagnose or predict future shortfalls more quickly. Rather than assuming a linear, additive set of force requirements based on the assumption that interventions are independent events, a revised planning framework would assume that the likelihood and frequency of military interventions in any year are affected not only by interests, risk, and presidential discretion but also by the number of interventions in recent years. Understanding that recent deployments may increase the likelihood of additional deployments in future years, planners can adjust force estimates to absorb these potential demands.

There are several ways quantitative metrics and estimates of temporal dependence, similar to those presented in the previous chapter, can be used directly to more accurately predict the frequency and timing of U.S. military interventions and to build force requirements that contain the appropriate numbers and types of personnel to absorb clustered demands. First, planners can use measures of the size and strength of the clustering effect to prepare for potential dependent clusters, once observing an increase in the number of military interventions in a given period. For example, the simple specifications in models 2 through 4 (Table 3.6) suggest that each additional intervention in one period increases the expected incidence of an intervention in the next by a factor of 1.25—a 25 percent increase over the base case. Other specifications suggest a stronger effect. This quantitative metric could be used as a guide in force planning. When the number of interventions in any year is higher than expected, even by a single intervention, planners can adjust force size or skill mix to reflect the higher risk for additional deployments in the next year, using the 1.25 expected increase as a foundation for the adjustments. The larger the increase in the number of interventions observed in previous periods, the greater the additional risk, and the more significant the revisions to force requirements would need to be.

Attempts to build temporal dependence into force planning may be most effective if planners consider exclusively (or nearly exclusively) interventions within a single geopolitical regime, since clustering seems to be sensitive to changes in the international system over time. Planners may also benefit from considering the nature of ongoing or recent interventions, focusing on the types of personnel who had been placed under the most strain during those deployments, since these are areas where additional demands from overlapping conflicts in the near future might go unmet. Attention to how far apart, in time, past interventions have occurred may also be useful for planners in understanding the potential implications of clustering for force requirements. Importantly, meeting these revised force requirements may necessitate some additional flexibility in both force planning and personnel management.

Another way that temporal dependence of military interventions could be incorporated quantitatively into force planning would be to use existing empirical specifications to develop vector error correction or vector autoregression models that could provide better insight into how changes in the number of interventions in one year will affect the number of interventions in subsequent years and even potentially the number of military personnel required.[2] Vector autoregression and vector error correction models were specifically developed to capture and describe interdependencies in time-series data, making them useful tools for the study of temporal dependence. These models are best interpreted using impulse-response functions, which show the change in the independent variable following a shock to the dependent variable and trace the effects of this shock over time. I did not use this kind of model here because it is less intuitive as a way to initially test for temporal dependence. However, impulse-response functions may ultimately be more useful for force planners than regression coefficients in informing decisions about how force requirements should be adjusted and when these adjustments should occur to properly account for temporal dependence. Planners could use this type of model to understand, for example, the change in number of interventions over a five-year period following a shock to interventions in the first year, as well as the timing of these additional interventions. This change in number of expected interventions could be used to project force requirements over the same period. A model that related number of deployments to force size, for example, could also be estimated using a vector error correction or autoregression approach. This would produce impulse-response functions able to estimate how an increase in numbers of deployments in one year might affect the size of the required force in the same and future years, accounting for temporal relationships.

Finally, temporal dependence can be included qualitatively in force planning processes, by incorporating an extended discussion of the concept and its implication, including how it relates to intervention duration, frequency, and concurrency, in planning documents. This additional discussion could include a description of the role played by geopolitical regime, the international and domestic characteristics that may affect the strength of temporally dependent clustering, and indicators planners can use to identify periods when temporally dependent clustering is more or less likely and shifts in the geopolitical dynamics that may affect clustering. This type of discussion could also be valuable in widening the discussion and awareness of temporal dependence and its implications among policymakers and force planners, providing additional support for the research agenda as a whole.

Avoiding Clustered Interventions

In addition to guiding force planning decisions, evidence that U.S. military interventions do seem to cluster at certain times may have some important implications or

---

[2] Vector autoregression is used to capture linear interdependencies in time-series data. Each variable is modeled as a function of its own lags and lags of other variables in the system of equations. Vector error correction models are a special type of vector autoregression used when time-series data are stationary, which means their mean and variance do not change over time.

lessons for strategic planners and individuals who make decisions about the deployment of U.S. forces in the future. Specifically, since clustered and overlapping interventions may place significant strain on military readiness, there are reasons that avoiding clusters altogether may be advantageous. Understanding the characteristics that make clustering more likely and the mechanisms through which clustering occurs may allow policymakers to intentionally avoid clusters in the future, either through revised policies and thresholds for the use of force or through additional institutional checks.

CHAPTER FIVE

Conclusion and Next Steps

This report provides initial evidence that temporal dependence does affect the timing of military interventions and contributes to deployments, although this result is sensitive to underlying geopolitical and prevailing strategic considerations and appears more likely under certain international regimes than others. When temporally dependent clustering occurs, each military intervention heightens the risk of another intervention in the near future. Importantly, even a small amount of temporal dependence between interventions could have a sizeable effect on the number of deployments and the demands placed on military personnel if this heightened risk aggregates over several periods or leads to multiple overlapping interventions. However, because temporal dependence is probabilistic and because resources are constrained, the growth of intervention clusters will not be infinite, and the risk for additional interventions will return to the mean over time. The report has also highlighted potential implications of temporal dependence for force planning and strategic decisions about the use of force and suggested some initial ways that temporal dependence might be addressed both quantitatively and qualitatively in planning and decisionmaking processes using the empirical results from Chapter Three.

The estimates and suggestions included here, however, are only a starting point. Additional data collection and empirical analyses could provide refined estimates of temporal dependence that are specific to particular types of deployments and could be used to disentangle the many possible mechanisms and drivers of temporally dependent clustering described in previous chapters. These types of estimates and insight are required to make the finding of temporal dependence actionable for policymakers. Producing these estimates would require at least three types of additional data. First, it will be necessary to investigate past intervention clusters in detail, looking at the types of activities involved in each, the timing and proximate trigger of each intervention, the evolution of each deployment, and the ways in which events in the deployment cluster were directly and indirectly related. This information will aid assessment of whether events during one military intervention directly contribute to the advent or necessity of the second. Additional data will also be needed on the domestic political processes that preceded and accompanied each deployment in the cluster. To the extent possible, this will include the political debates over the intervention, the implicit and stated rationales for the deployment, any existing opposition arguments, and any other political or economic dynamics that might have encouraged the deployment. The statements U.S. political and military leaders made before each intervention in a cluster may also be important, since they may reveal or create commitments that lock in future interventions Finally, it will be necessary to explore the systemic political and economic situation at the time of the intervention clustering, including the distribution of power, the relative status of the United States and domestic perceptions of this status, and the trend in U.S. relative power leading up to the intervention. This would include identifying suitable proxy variables for characteristics of the geopolitical regime, which may explain its strong relationship with the likelihood and strength of clustering between military interventions.

In addition to the collection of additional data that refines the coding of interventions, there are several other ways to improve the understanding and estimates of temporal dependence in this report. First, existing estimates of temporal dependence can be refined with the inclusion of some additional controls, including more-sensitive measures of global economic conditions and of U.S. capabilities. For example, variables that measure U.S. interests as represented in United Nations votes relative to allies, competitors, countries in which it might intervene, and vote outcomes could provide a better relative measure of U.S. goals and status than the CINC score alone. It also may be valuable to assess whether a more disaggregated measure of political instability that considers the effects of intrastate conflict, interstate disputes, and political instability (riots, coups) separately is more strongly related to intervention timing. Second, as described above, vector autoregression and vector error correction models can be used to provide more useful insight into how changes in number of interventions in one period (or other variables, such as instances of instability or economic growth) affect the number of interventions over subsequent periods. Impulse-response functions would provide insight into the timing, duration, and magnitude of any temporal dependence that may contribute to more-sensitive force planning decisions. A third refinement to the existing work would be to provide a more disaggregated estimate of temporal dependence that considers deployment type. This will require some additional qualitative research to properly define relevant types of interventions and to assign interventions in the data set to these categories. Similarly, it may be possible to explore region-specific clusters that would tie this work to that focused on regional clusters in instability and conflict.

Bibliography

Alesina, Alberto, and Roberto Perotti, "Income Distribution, Political Instability, and Investment," Cambridge, Mass.: National Bureau of Economic Research, Working Paper No. 4486, 1993.

Allen, Franklin, and Douglas Gale, "Financial Contagion," *Journal of Political Economy*, Vol. 108, No. 1, February 2000, pp. 1–33.

Bae, Kee-Hong, G. Andrew Karolyi, and René M. Stulz, "A New Approach to Measuring Financial Contagion," *Review of Financial Studies* Vol. 16, No. 3, Autumn 2003, pp. 717–763.

Beck, Nathaniel, Jonathan N. Katz, and Richard Tucker, "Taking Time Seriously: Time-Series-Cross-Section Analysis with a Binary Dependent Variable," *American Journal of Political Science*, Vol. 42, No. 4, October 1998, pp. 1260–1288.

Bercovitch, J., "Conflict Management of Enduring Rivalries: The Frequency, Timing, and Short-Term Impact of Mediation," *International Interactions*, Vol. 22, No. 4, 1997.

Berry, Nicholas, Marcus Corbin, Christopher Hellman, Jeffrey Mason, Daniel Smith, Rachel Stohl, and Tomas Valaseck, "U.S. Military Deployments/Engagements, 1975–2001," *Military Almanac 2001–2002,* Washington, D.C.: Center for Defense Information, November 2001.

Brands, H. W., *Cold Warriors: Eisenhower's Generation and American Foreign Policy,* New York: Columbia University Press, 1988.

Bueno de Mesquita, Bruce, Alastair Smith, Randolph Siverson, and James D. Morrow, *Logic of Political Survival*, Cambridge, Mass.: MIT Press, 2003.

Buhaug, Halvard, and Kristian Skrede Gleditsch, "Contagion or Confusion? Why Conflicts Cluster in Space," *International Studies Quarterly* Vol. 52, No. 2, June 2008, pp. 215–233.

Carmazza, Francesco, Luca Antonio Ricci, and Ranil Salgado, "Trade and Financial Contagion in Currency Crises," working paper, Washington D.C.: International Monetary Fund, March 2000.

Center for Systemic Peace, Integrated Network for Societal Conflict Research, "Polity IV: Regime Authority Characteristics and Transitions Datasets," website, 1800–2010. As of July 23, 2012: http://www.systemicpeace.org/inscr/inscr.htm

Collier, Paul, and Anke Hoeffler, "On the Incidence of Civil War in Africa," *Journal of Conflict Resolution*, Vol. 46, No. 1, February 2002, pp. 13–28.

———, "Resource Rents, Governance, and Conflict," *Journal of Conflict Resolution*, Vol. 49, No. 4, August 2005, pp. 625–633.

Correlates of War Project, "Militarized Interstate Disputes v3.10," database, 2007. As of September 1, 2011: http://www.correlatesofwar.org/COW2%20Data/MIDs/MID310.html

————, "Correlates of War Project National Material Capabilities Data Documentation," Version 4.0, June 2010.

Corsetti, Giancarlo, Marcello Pericoli, and Massimo Sbracia, "Some Contagion, Some Interdependence: More Pitfalls in Tests of Financial Contagion," *Journal of International Money and Finance*, Vol. 24, No. 8, December 2005, pp. 1177–1199.

Fearon, James, "Domestic Political Audiences and the Escalation of International Disputes," *American Political Science Review*, Vol. 88, No. 3, September 1994, pp. 577–592.

Fearon, James, and David Latin, "Ethnicity, Insurgency, and Civil War," *American Political Science Review*, Vol. 97, 2003, pp. 75–90.

Gleditsch, Nils Petter, Peter Wallensteen, Mikael Eriksson, Margareta Sollenberg, and Håvard Strand, "Armed Conflict 1946–2001: A New Dataset," *Journal of Peace Research*, Vol. 39, No. 5, September 2002.

Goldstone, Jack, "Population and Security: How Demographic Change Can Lead to Violent Conflict," *Journal of International Affairs*, Vol. 56, 2002.

Goldstone, Jack A., Robert H. Bates, David L. Epstein, Ted Robert Gurr, Michael B. Lustik, Monty G. Marshall, Jay Ulfelder, and Mark Woodward, "A Global Model for Forecasting Political Instability," *American Journal of Political Science*, Vol. 54, No. 1, January 2010, pp. 190–208.

Grimmett, Richard, *U.S. Use of Preemptive Military Force*, Washington, D.C.: Congressional Research Service, September 2002. As of September 1, 2011:
http://fpc.state.gov/documents/organization/13841.pdf

————, *The War Powers Resolution: After Thirty-Six Years*, Washington, D.C.: Congressional Research Service, April 2010. As of September 1, 2011:
http://www.fas.org/sgp/crs/natsec/R41199.pdf

Huang, Haizhou, and Chenggang Xu, "Financial Institutions, Financial Contagion, and Financial Crises," working paper, Washington, D.C.: International Monetary Fund, 2000.

Huntington, Samuel, *Political Order in Changing Societies*, New Haven, Conn.: Yale University, 1968.

International Monetary Fund, "World Economic Outlook Databases," website, various dates. As of July 23, 2012:
http://www.imf.org/external/ns/cs.aspx?id=28

James, Patrick, and John O'Neal, "The Influence of Domestic and International Politics on the President's Use of Force," *Journal of Conflict Resolution*, Vol. 35, No. 2, 1991, pp. 307–332.

Kanter, Arnold, and Linton Brooks, *U.S. Intervention Policy for the Post Cold War World: New Challenges and New Responses,* New York: Norton, 1994.

Kaufmann, Chaim, "Intervention in Ethnic and Ideological Civil Wars: Why One Can Be Done the Other Can't," *Security Studies*, Vol. 6, No. 1, 1996, pp. 62–103.

Klare, Michael T., *Beyond the "Vietnam Syndrome": U.S. Interventionism in the 1980s*, Washington, D.C.: Institute for Policy Studies, 1981.

Kodres, Laura, and Matthew Pritsker, "A Rational Expectations Model of Financial Contagion," *Journal of Finance*, Vol. 57, No. 2, April 2002, pp. 769–799.

Lian, Bradley, and John R. Oneal, "Presidents, the Use of Military Force, and Public Opinion," *Journal of Conflict Resolution*, Vol. 37, No. 2, June 1993.

Marte, Ana, and Winslow Wheeler, eds., *CDI Military Almanac*, Washington, D.C.: Center for Defense Information, 2007.

Mansfield, Edward, "The Distribution of Wars over Time," *World Politics*, Vol. 41, No. 1, October 1988, pp. 21–51.

Meernik, James, "Presidential Decision Making and the Political Use of Military Force," *International Studies Quarterly*, Vol. 38, No. 1, March 1994, pp. 121–138.

Ostrom, Charles W., Jr., and Brian L. Job, "The President and the Political Use of Force," *American Political Science Review*, Vol. 80, No. 2, June 1986, pp. 541–566.

Pearson, Frederic S., and Robert A. Baumann, "Foreign Military Intervention and Changes in United States Business Activity," *Journal of Political and Military Sociology*, Vol. 5, No. 1, Spring 1977, pp. 79–97.

Political Instability Task Force, "PITF Phase V Findings (through 2004)," website, 2005. As of February 8, 2013:
http://globalpolicy.gmu.edu/political-instability-task-force-home/pitf-phase-v-findings-through-2004/

Pollins, Brian, "Global Political Order, Economic Change, and Armed Conflict: Coevolving Systems and the Use of Force," *American Political Science Review*, Vol. 90, No. 1, March 1996, pp. 103–117.

Regan, Patrick M, "Choosing to Intervene: Outside Interventions in Internal Conflicts," *Journal of Politics*, Vol. 60, No. 3, August 1998, pp. 754–779.

Regan, Patrick M., and Aysegul Aydin, "Diplomacy and Other Forms of Intervention in Civil Wars," *Journal of Conflict Resolution*, Vol. 50, No. 5, October 2006, pp. 736–756.

Regan, Patrick M., and Allan C. Stam, "In the Nick of Time: Conflict Management, Mediation Timing, and the Duration of Interstate Disputes," *International Studies Quarterly*, Vol. 44, No. 2, June 2000, pp. 239–260.

Sambanis, Nicholas, "Do Ethnic and Nonethnic Civil Wars Have the Same Causes? A Theoretical and Empirical Inquiry (Part 1)," *Journal of Conflict Resolution*, Vol. 45 No. 3, June 2001, pp. 259–282.

———, *Understanding Civil War: Evidence and Analysis,* Washington, D.C.: World Bank, 2005.

Singer, J. David, Stuart Bremer, and John Stuckey, "Capability Distribution, Uncertainty, and Major Power War, 1820–1965," in Bruce Russett, ed., *Peace, War, and Numbers*, Beverly Hills: Sage, 1972, pp. 19–48.

Siverson, Randolph M., and Michael P. Sullivan, "The Distribution of Power and the Onset of War," *Journal of Conflict Resolution*, Vol. 27, No. 3, September 1983, pp. 473–494.

Sullivan, Patricia, and Michael Koch, "Military Intervention by Powerful States, 1945–2003," *Journal of Peace Research*, Vol. 46, No. 5, September 2009, pp. 707–718.

Uppsala University, Department of Peace and Conflict Resolution, "UCDP/PRIO Armed Conflict Dataset v.4-2012," database, Uppsala, Sweden, 1946–2011. As of July 20, 2012:
http://www.pcr.uu.se/research/ucdp/datasets/ucdp_prio_armed_conflict_dataset/

United Nations Statistics Division, "International Merchandise Trade Statistics," website, 2013. As of July 23, 2012:
http://unstats.un.org/unsd/trade/imts/imts_default.htm

Urdal, Henrik, "A Clash of Generations? Youth Bulges and Political Violence," *International Studies Quarterly,* Vol. 50, No. 3, September 2006, pp. 607–629.

Ward, Michael, and Kristian Skrede Gleditsch, "Location, Location, Location: An MCMC Approach to Modeling the Spatial Context of War and Peace," *Political Analysis,* Vol. 10, No. 3, 2000, pp. 244–260.

Yoon, Mi Yung, "Explaining U.S. Intervention in Third World Internal Wars, 1945–1989," *Journal of Conflict Resolution*, Vol. 41, No. 4, August 1997, pp. 580–602.

Zartman, I. W., *Ripeness: The Hurting Stalemate and Beyond,* Washington, D.C.: National Academies Press, 2000.